W9-AFD-524

NORTH-HOLLAND
MATHEMATICS STUDIES 11

Notas de Matemática (52)
Editor: Leopoldo Nachbin
Universidade Federal do Rio de Janeiro
and University of Rochester

Analytic Functions and Manifolds in Infinite Dimensional Spaces

G. COEURÉ

Université de Nancy I

1974

NORTH-HOLLAND PUBLISHING COMPANY – AMSTERDAM • LONDON
AMERICAN ELSEVIER PUBLISHING COMPANY, INC. – NEW YORK

© NORTH-HOLLAND PUBLISHING COMPANY – 1974

All Rights Reserved. No part of this publication may be reproduced, stored in a retrieval system or transmitted, in any form or by any means, electronic, mechanical, photocopying, recording or otherwise, without the prior permission of the Copyright owner.

Library of Congress Catalog Card Number: 73 93562

ISBN North-Holland:

Series: 0 7204 2700 2

Volume: 0 7204 2711 8

ISBN American Elsevier: 0 444 10621 9

PUBLISHERS:

NORTH-HOLLAND PUBLISHING COMPANY – AMSTERDAM

NORTH-HOLLAND PUBLISHING COMPANY, LTD. – LONDON

SOLE DISTRIBUTORS FOR THE U.S.A. AND CANADA:

AMERICAN ELSEVIER PUBLISHING COMPANY, INC.

52 VANDERBILT AVENUE

NEW YORK, N.Y. 10017

PRINTED IN THE NETHERLANDS

MATH.-SCI.
QA 331
.C64
1407907
Math Sci
Sep.

Preface

This book is written from a course given by author at the Federal University of Rio de Janeiro during the spring and summer quarters 1972. The aim has not been to write a complete review in any direction of works about infinite dimensional complex analysis, but to provide a systematic approach for analytic continuation of analytic mappings in infinitely many variables. For this reason many interesting results are not developed here, nevertheless the author hopes that the bibliography is almost everywhere complete.

According to a way of H. Cartan [15] and K. Stein [82] used by M. Schottenloher [79] , chapter I develops the properties of spread manifolds endowed with an analytic sheaf which are needed.

Chapter III starts with the basic properties of analytic mappings. They are shortly proved, they could be found in M. Hervé's [40] and Ph. Noverraz's books [70] . The second section gives somme examples of analytic continuations which show new facts arising from infinite dimensions. For instance, the classical Riemann extension theorem and analytic continuation in a product $\mathbb{C}^I$.

In chapter II, IV, V, we develop the properties concerning the simultaneous continuation of some natural Fréchet spaces of analytic mappings which generalise some earlier results about functions of bounded [25] and F_τ [19,60] type, and regular classes [79] . Theorems of Cartan-Thullen type are obtained and also the holomorphic convexity of their envelope of holomorphy when the underlying space satisfies a Grothendieck's approximation property. Chapter II takes place before because no vector structure on the underlying space is useful in this chapter. Chapter IV provides an imbedding for the maximal extension of some analytic algebra in its spectrum for suitable topologies. The method is successful when the underlying space E is a metrizable locally convex vector space. But the indentification with the spectrum is an open problem which is solved when E is a product $\mathbb{C}^I$.

Chapter VII is mainly concerned with the existence of the following commutative diagram.

$$\begin{array}{ccc} X & \xrightarrow{\varphi} & Y \\ \downarrow & & \downarrow \\ \tilde{X} & \xrightarrow{\tilde{\varphi}} & \tilde{Y}(A) \end{array}$$

Here X and Y are spread manifolds, $\tilde{X}$ is the maximal extension of X for the whole set of analytic functions, $\tilde{Y}(A)$ the maximal entension of Y for an algebra of analytic functions and φ is a given analytic mapping from X to Y. The general problem is still open, but the used methods are successful when Y is a sequentially complete locally convex vector space, or φ is a local isomophisme, or A a natural Fréchet algebra.

Chapter VIII mainly generalises the Runge theorem in finite dimensional Stein manifolds X toward some particular open sets of the product of X with a complex locally convex topological space.

I am greatly indebted to Professor Leopoldo Nachbin who has encouraged the improvement of the original notes during and after my visiting time in Brasil. Thanks to him, this book can appear in his collection "Notas de Mathemàtica".

I am also indebted to M. Hamadi for her typing work.

Gérard Coeuré

Federal University of Rio de Janeiro
July 1972
and
University of Nancy I
Septembre 1973

CONTENTS

CHAPTER I : SPREAD MANIFOLDS

§ 1.- Intersection of morphisms.

In this section, E is a Hausdorff locally connected space. A manifold spread over E is a pair (X,p) with X a Hausdorff space and p a map from X into E which is a local homeomorphism. Every connected component of X is open and spread over E by p, so we shall assume that X is connected ; we shall say "manifold" for "manifold spread over E". We shall denote p_x^{-1} as the inverse of p which is locally defined in a neighbourhood of $p(x)$ by $p_x^{-1} \circ p =$ identity.

Proposition 1.1.- A subspace ω of X, homeomorphic with a domain V of E by p is an open and connected component of $p^{-1}(V)$.

Proof. Let Ω be the connected component of $p^{-1}(V)$ which contains ω and suppose $\omega \neq \Omega$. Then a boundary point x_o of ω as subspace of Ω verifies $x_o = (p|_\omega)^{-1}\,[p(x_o)]$ using a net which converges to x_o in ω, but $(p|_\omega)^{-1}\,[p(x_o)]$ cannot belong to a neighbourhood of x_o where the restriction of p is a homeomorphism by using a net with converge to x_o in $\Omega - \omega$.

Corollary.- Each fiber $p^{-1}(a)$ is discrete.

Proposition 1.2.- If there is a countable basis B for the open sets of E, then X has also a countable basis such that every open set of this basis is homeomorphic by p with a domain in E.

Proof. Since E is locally connected, we can assume that sets which belong to B are domains ; we denote by $\tilde{B}$ the set of domains in X which are holeomorphic by p with a domain of B ; let X_m be the set of points x, $x \in X$, which can be joined to some x_o by a chain of m domains belonging to $\tilde{B}$. Each X_m is open, and X is the union of X_m since X is connected, therefore we have just to prove the property for each X_m.

By proposition 1.1, $X_1 = \bigcup_\omega p_{x_o}^{-1}(\tilde{\omega})$, with $\omega = p(\tilde{\omega})$, $\tilde{\omega} \in \tilde{B}^m$, and $x_o \in \tilde{\omega}$; the set of such ω is countable and $p_{x_o}^{-1}(\tilde{\omega})$ verifies the property, so it is proved for X_1.

Now, we assume that X_m verifies the property and let $(\tilde{\omega}_n)$ be a convenient basis of X_m ; the union of X_m with $\{\tilde{\omega} \in \tilde{B} \mid \exists\, \tilde{\omega}_n \text{ with } \tilde{\omega}_n \subset \tilde{\omega}\}$ contains X_{m+1} ; by proposition 1.1, there is only one $\tilde{\omega} \in \tilde{B}$ such that $\tilde{\omega}$ contains some domain $\tilde{\omega}_n$.
Therefore X_{m+1} is a countable union of domains with a convenient basis, and X_{m+1} has also a convenient basis.

<u>Corollary</u>.- *If E has a countable basis of open sets, then each fiber $p^{-1}(a)$ is countable or finite.*

<u>Definition</u> 1.1.- *A map u from X to the space X' of an other manifold (X',p') is a morphism if u is continuous and satisfies $p = p' \circ u$. A morphism is an isomorphism if it is one-to-one.*

<u>Proposition</u> 1.3.- *A morphism is always open, and two different morphisms from X to X' satisfy : $u(x) \neq v(x)$ for all $x \in X$.*

<u>Proof</u>. Clear by connectedness.

<u>Theorem</u> 1.1.- *Let u be a family of morphisms from X to (X_i,p_i). There exists a manifold denoted by $\cap X_i$, a morphism from X to $\cap X_i$ denoted by $\cap u_i$, morphism φ_i from $\cap X_i$ to X_i , such that :*

(i) $\varphi_i \circ (\cap u_i) = u_i$ for any i .

(ii) For every manifold (X',p') , and morphism u' from X to X' , and morphisms u_i' from X to X_i such that $u_i = u_i' \circ u$ for any i, there exists a morphism φ' from X' to $\cap X_i$ such that : $\cap u_i = \varphi' \circ u$ and $u_i' = \varphi_i \circ \varphi'$.

All the manifolds $\cap X_i$ which verify the above properties are isomorphic. The morphism $\cap u_i$ will be called the intersection of morphisms u_i .

<u>Proof</u>. In the product space $\prod X_i$, we consider the set Y of points $\bar{x} = (x_i)$ for which the projections $p_i(x_i)$ are one point $\bar{p}(\bar{x})$ and there exists a neighbourhood V of $\bar{p}(\bar{x})$ which is homeomorphic with a neighbourhood ω_i of x_i by p_i for each index i . From such ω_i , we define a basis $\prod \omega_i \cap Y$ for a Hausdorff topology on Y and so $(Y,\bar{p})$ is spread over E.

Let $\cap u_i$ be the map from X into $\prod X_i$ defined by : $(\cap u_i)(x) = (u_i(x))$. Since $p_i \circ u_i = p$, there exists a neighbourhood ω of x such that $u_i(\omega)$ is homeomorphic with $p(\omega)$ by p_i and therefore $\bigcap u_i$ is a morphism from X to Y. We take for $\bigcap X_i$ the connected component of $(\bigcap u_i)(X)$ in Y, φ_i is the canonical projection of $\prod X_i$ onto X_i and the map φ' is the one defined by $\varphi'(x') = (u_i'(x'))$. It satisfies $\varphi' \circ u' = \bigcap u_i$, therefore $\varphi'(X')$ contains $(\bigcap u_i)(X)$, is connected and $\varphi'(X')$ is included in $\bigcap X_i$.

Only uniqueness remains to be proved. Let X_1 and X_2 be two intersections with the properties of previous theorem, we denote by u_1 and u_2 the associated intersection of morphisms u_i. We can particularize theorem 1.1 in the following way : $u' = u_1$, $u_i' = \varphi_i$, $X' = X_1$, $\bigcap u_i = u_2$, $\bigcap X_i = X_2$. From (ii) we get $u_2 = \varphi_2' \circ u_1$ where φ_2' is a morphism from X_1 to X_2 and we also have $u_1 = \varphi_1' \circ u_2$ with φ_1' from X_2 to X_1. Checking up these relations, we can easily verify that φ_1' is an isomorphism.

§ 2.- Maximal extensions.

Let Z be another locally connected Hausdorff space and $F_E(Z)$ a subpresheaf of sets of continuous functions defined on E with values in Z. The set of sections over an open set U in E will be noted $F_E(U,Z)$, and the germ at x of $f \in F_E(U,Z)$ as f_x. In the following sections, we shall assume the following analytic property of $F_E(Z)$:

(A) : For every domain U in E, f and g in $F_E(U,Z)$, $x_0 \in U$, $f_{x_0} = g_{x_0}$ imply $f = g$.

Let (X,p) and (Y,π) be two manifolds spread over E and Z. We introduce the presheaf $F_X(Y)$ whose sections over any open set U in X are continuous, Y-valued, and satisfy : $(\pi \circ f \circ p_x^{-1})_{p(x)} \in F_E(Z)$ for all $x \in U$.

Proposition 1.4.- For every $u \in F_E(E,Z)$, $u \circ p \in F_X(X,Z)$.

Proposition 1.5.- Given a morphism u from X to X', then the transpose mapping $u^* : f \to f \circ u$ maps $F_{X'}(X',Y)$ into $F_X(X,Y)$.

Proposition 1.6.- The presheaf $F_X(Y)$ satisfies property (A).

The above propositions are obvious. Let now (Y_i,π_i) be a family of manifolds spread over Z_i, $F_E(Z_i)$ a family of presheafs with property

(A) , which is indexed by the same set.

Definition 1.2.- Let $\Gamma = f_i$ be a family with $f_i \in F_X(X,Z_i)$. A morphism u from X to X' is called a Γ-extension for $F_E(Z_i)$, if there exists $f'_i \in F_{X'}(X',Z_i)$ with $f_i = f'_i \circ u$ for each index i . The family $\Gamma' = (f'_i)$ is unique because of (A) , and is called the extension of Γ to X' and denoted by $u^*(\Gamma)$.

Definition 1.3.- A Γ-extension $u : X \to X'$ is called maximal if every Γ-extension $v : X \to X'$ can be factored through a morphism $w : X'' \to X'$, (that is $u = w \circ v$) .

Proposition 1.7.- All Γ-maximal extensions are isomorphic and are $v^*(\Gamma)$-extensions of any other Γ-extension v .

Proof. If the previous u and v are maximal, there exists a morphism φ from X' to X'' such that : $u = w \circ \varphi \circ u$ and $v = \varphi \circ w \circ v$, By proposition 1.3 $w \circ \varphi$ and $\varphi \circ w$ are the identity and therefore w is an isomorphism. On the other hand w is a $v^*(\Gamma)$-extension.

Theorem 1.2.- We assume that the previous family is the union sub-families Γ_j . Let $u_j : X \to X_j$ be a maximal Γ_j-extension, then $\cap u_j$ is a maximal Γ-extension.

Proof. We recall the existence of morphisms $\varphi_j : \cap X_j \to X_j$ with $u_j = \varphi_j \circ (\cap u_j)$. Every $f \in \Gamma$ has an extension f_j to X_j , therefore $f_j \circ \varphi_j$ is an extension to $\cap X_j$.

Let us now prove the maximality of $\cap X_j$. Given $u' : X \to X'$ a Γ-extension, the restriction of u' to Γ_j can be factoried through a morphism $u'_j : X \to X_j$ with $u_j = u'_j \circ u'$. We can apply (ii) of theorem 1.1 and the morphism φ' of (ii) gives the maximality.

Theorem 1.3.- There exists a maximal Γ-extension.

Proof. By the previous theorem, we have only to construct the maximal extension of each function $f \in F_E(X,Y)$. First we endow $F_E(Y)$ with a structure of a manifold in the following way. For every $f \in F_E(U,Y)$, U domain in E , we consider $N(h,U) = \{h_x \mid x \in U\}$. It can be seen without difficulty that we have built a basis for a Hausdorff topology on $F_E(Y)$, and $\bar{p} : h_x \to x$ is a homeomorphism from N(h,U) onto U , Further,

$u : x \to (f \circ p_x^{-1})_{p(x)}$ is a morphism from X to $(F_E(Y), \bar{p})$. Let $\bar{X}$ be the connected component of $u(X)$ into $F_E(Y)$; $\bar{X}$ is a f-extension, actually the map $\bar{f} : h_x \to h_x(x)$ from X into Y is clearly continuous, satisfies $\bar{f} \circ (\bar{p}_{h_x})^{-1} = h$ on U for all $h \in F_E(U,Y)$ and $x \in U$, therefore $\bar{f}$ is a section of $F_{\bar{X}}(Y)$ over X and $\bar{f} \circ u = f$.

We shall now prove the maximality of $\bar{X}$. Let $v : X \to X'$ be a f-extension with f' the extension of f to X' . By the next computation it is clear that $x' \to (f' \circ p'^{-1}_x)_{p'(x)}$ is a morphism from X to $\bar{X}$ which factors u through v. We have :

$$\left[f' \circ (p')^{-1}_{v(x)}\right]_{p' \circ v(x)} = (f' \circ v \circ p_x^{-1})_{p(x)} = (f \circ p_x^{-1})_{p(x)} = u(x) \text{ .}$$

§ 3.- Separation properties.

We shall apply the previous theorem to a subset Γ in $F_E(X,Y)$. In what follows, $\bar{X}$ is the Γ-maximal extension of X for $F_E(Y)$, and $\bar{\Gamma}$ is the extension of Γ to $\bar{X}$.

Proposition 1.8.- Whenever Γ *separates* X *and the morphism* u *from* X *to* $(\bar{X},\bar{p})$ *is injective then* X *is isomorphic with the domain* $u(X)$ *in* $\bar{X}$.

Proof. Clear by the relation $\bar{i} = f \circ u$.

Proposition 1.9.- If Γ *contains a set of functions* $g \circ p$, $g \in F_E(E,Y)$ *and* E *is separated by the function* g *then* $\bar{X}$ *is separated by* $\bar{\Gamma}$.

Proof. Let us denote by X_s the separated quotient space of X associated with the relation : $\bar{f}(x) = \bar{f}(x')$, all $\bar{f} \in \bar{\Gamma}$. Since the functions $g \circ \bar{p}$ are in $\bar{\Gamma}$, all the points of some class in $\bar{X}_s$ have the same projection on E . Then the induced mapping $\bar{p}$ on $\bar{X}_s$ is a local homeomorphism, and the quotient map from $\bar{X}$ to $\bar{X}_s$ is a morphism and a $\bar{\Gamma}$ extension. By proposition 1.7, $\bar{X}$ and $\bar{X}_s$ are isomorphic.

§ 4.- Univalent extensions.

X is a domain in E and Γ a subset of $F_E(X,Y)$ which separates X .

Proposition 1.10.- The following properties are equivalent :

(i) There is no pair (U,V) , U *and* V *domains in* E ,

$U \subset V \cap X$, $V - X \neq \emptyset$, such that for every $f \in \Gamma$ there exists $\bar{f} \in F_E(V,Y)$ such that $\bar{f}|_U = f|_U$.

(ii) X is isomorphic with its maximal Γ-extension.

Proof. (i) $\Longrightarrow$ (ii) : Let $\bar{X}$ be the maximal Γ-extension of X , with its injective morphism $u : X \to \bar{X}$. We must prove $u(X) = \bar{X}$. If x_o is a boundary point of $u(X)$, then $\bar{p}(x_o)$ is a boundary point of X . Let W be a connected neighbourhood of x_o with $\bar{f} \circ \bar{p}_{x_o}^{-1} \in F_E(W),Y)$ for every $\bar{f} \in \Gamma$. Then $\bar{f} \circ p_{x_o}^{-1}(a) = f(a)$ for all $a \in \bar{p}(W) \cap u^{-1}(W)$ and the pair $(U,\bar{p}(W))$ is like (i) where U is a domain in $\bar{p}(W) \cap u^{-1}(W)$.

(ii) $\Longrightarrow$ (i) : Given a pair (U,V) like (i), the mapping $x \to (\bar{f}_x)_{f \in \Gamma}$ is a morphism from V to $\bar{X}$. Since $V - X \neq \emptyset$, $\bar{X}$ cannot be isomorphic with X .

CHAPTER II : NATURAL FRECHET SPACES

Here, Z is a locally convex Frechet space whose set of semi-norms that defines its topology is denoted by $N(Z)$; E is a Hausdorff locally connected space ; $C_E(Z)$ is the presheaf of continuous and Z-valued functions ; $C_E^\infty(U,Z)$ is the subspace of $C_E(U,Z)$ whose functions are bounded, equipped with the uniform topology.

Throughout this chapter, $F_E(Z)$ is a sub-presheaf of $C_E(Z)$ which satisfies (A) and the next property (B).

(B) $F_E(U,Z) \cap C_E^\infty(U,Z)$ is closed in $C_E^\infty(U,Z)$, for all open sets U in E.

In what follows, X is a manifold spread over E.

Definition 2.1.- *Given Γ a set in $F_X(X,Z)$; a part T of X will be called a bounding set for Γ, iff $\|q \circ f\|_T < \infty$ all $f \in \Gamma$, $q \in N(Z)$. We are using the notation $\|.\|_T$ for $\sup_{x \in T} (.)$.*

Definition 2.2.- *This set Γ is called locally bounded iff, for all $x \in X$, each $f \in \Gamma$ is bounded in a neighbourhood ω of x ; we say Γ is uniformly bounded whenever the previous ω can be chosen independantly of f .*

Remark. When Z is normed, $F_X(X,Z)$ is always locally bounded.

Definition 2.3.- *A vector subspace Γ of $F_X(X,Z)$, with a locally convex linear topology $\mathcal{C}$ is called natural whenever $\mathcal{C}$ is stronger than the topology of pointwise convergence. The set of semi-norms for $\mathcal{C}$ is denoted by $N(\Gamma)$.*

Proposition 2.1.- *Given Γ a natural Frechet space in $F_X(X,Z)$ and T a bounding set for Γ , then $f \to \|q \circ f\|_T$ is a continuous semi-norm of Γ, for each $q \in N(Z)$.*

Proof. The set $\{f \mid \|q \circ f\|_T \leqslant 1\}$ is a barrel in Γ , which implies the continuity of the semi-norm since Γ is barreled. Actually this previous set is closed because of continuity of the evaluation at any x in X ; it is absorbing set because of the boundness property of T and it is obviously balanced.

Corollary 2.1.- A natural Frechet space has a stronger topology than the compact open topology.

Proof. Clear, with T as a compact set.

Definition 2.4.- Let $\mathcal{U}$ be a covering of X by open sets, and denote by $A_{\mathcal{U}}$ the set of $f \in F_X(X,Z)$ such that $f(\omega)$ is bounded for every $\omega \in \mathcal{U}$. We endow $A_{\mathcal{U}}$ with the topology defined by the semi-norms :
$\|f\|_{\omega,q} = \|q \circ f\|_{\omega}$, $\omega \in \mathcal{U}$, $q \in N(Z)$.

Proposition 2.2.- Whenever $\mathcal{U}$ is countable, $A_{\mathcal{U}}$ is an uniformly bounded, Frechet space.

Proof. Clear, using (B).

Proposition 2.3.- Let Γ be a natural Frechet space. Given $x \in X$ and ω as a domain in E with $p(x) \in \omega$. Let $\Gamma_{\omega,x}$ be defined as :
$\{f \in \Gamma \mid \exists\ \tilde{f} \in F_E(\omega,Z)\ ,\ \tilde{f}(\omega)$ *bounded, and* $(f \circ p_x^{-1})_{p(x)} = \tilde{f}_x\}$. *Then, either $\Gamma = \Gamma_{\omega,x}$ or $\Gamma_{\omega,x}$ is a set of first category in Γ.*

Proof. The map $\tilde{f}$ is unique because of property (A), and therefore the set of pairs $(f,\tilde{f})$, $f \in \Gamma_{\omega,x}$, is a vector subspace of $\Gamma \times F_E(\omega,Z)$ denoted by $\tilde{\Gamma}_{\omega}$.

We claim now that semi-norms $\|(f,\tilde{f})\|_{q,\pi} = \pi(f) + \|\tilde{f}\|_{\omega,q}$, $\pi \in N(\Gamma)$ and $q \in N(Z)$, define a Frechet topology on $\tilde{\Gamma}_{\omega}$. Actually, given a Cauchy sequence $(f_n,\tilde{f}_n)$ in $\tilde{\Gamma}_{\omega}$, f_n converges near some $f \in \Gamma$ and $\tilde{f}_n$ near some $\tilde{f} \in F_E(\omega,Z)$ by property (B) , and $\tilde{f}(\omega)$ is bounded. Furthermore $f_n \circ p_x^{-1} = \tilde{f}_n$ in a domain ω' , such that $p(x) \in \omega'$, $\omega' \subset \omega$, and p_x^{-1} defined in ω' . Since the topology of Γ is stronger than the point-wise topology, we have $f \circ p_x^{-1} = \tilde{f}$ in ω' ; hence $(f,\tilde{f}) \in \tilde{\Gamma}_{\omega}$.

Finally, the map $(f,\tilde{f}) \to f$, from $\tilde{\Gamma}_{\omega}$ to Γ , is linear and continuous, $\Gamma_{\omega,x}$ is its range. Then the stated proposition is now a consequence of a well known Banach theorem.

Theorem 2.1.- If each point of E has a countable basis of neighbourhood, then every locally bounded natural Frechet space is uniformly bounded.

Proof. Let $x \in X$ be given and let (ω_n) be such a set of neighbourhoods of $p(x)$. If Γ were not uniformly bounded, we should have

$\Gamma_{\omega_{n,x}} \neq \Gamma$ for all n and there would exist some $f \in \Gamma - \bigcup \Gamma_{\omega_n}$ by proposition 2.3 and the Baire property of Γ ; this is impossible since $(f \circ p_x^{-1})(\omega_n)$ is bounded for some n .

Theorem 2.2.- *If Z is normed and E has the countability property above, then every natural Frechet space Γ can be continuously imbedded into a convenient class $A_{\mathcal{U}}$ with $\mathcal{U}$ countable.*

Proof. Since Z is normed, Γ is locally bounded and therefore uniformly bounded by the previous theorem ; there exists an open set ω in X such that $f(\omega)$ is bounded in Z for every $f \in \Gamma$. In the following argument we denote by $\|\cdot\|$ the norm of Z and by $(\pi_n) = N(\Gamma)$. We consider the set X_n defined as :

$$X_n = \{x \in X \mid \|f(x)\| \leqslant n . [\pi_n(f) + \|f\|_\omega] \quad , \text{ all } f \in \Gamma \} .$$

We claim that $X = \bigcup \mathring{X}_n$, which is enough to construct $\mathcal{U}$. Actually, if that is not true, there would exist $x \in X$ and a sequence (x_n) which converges near x with $x_n \in X - X_n$ for all n . For each n there would exists $f_n \in \Gamma$ with $\|f_n(x_n)\| > n . [\pi_n(f_n) + \|f_n\|_\omega]$; f_n must be $\neq 0$ and therefore $\|f_n\|_\omega \neq 0$.
Using $g_n = [\pi_n(f_n) + \|f_n\|_\omega]^{-1} . f_n$, we should have :

$$\|g_n(x_n)\| > n \quad , \quad \pi_n(g_n) \leqslant 1$$; then g_n is a bounded sequence which must be bounded on the relatively compact set (x_n) by proposition 2.1.

To prove the continuity of the imbedding map $\Gamma \to A_{\mathcal{U}}$, we check the continuity of $f \to \|f\|_\omega$ as in proposition 2.1, with ω any bounding set for Γ .

Proposition 2.4.- *Let Γ be a natural Frechet space which separates X ; given two points x and x' in X , $x \neq x'$, then the set $\{f \in \Gamma \mid f(x) \neq f(x')\}$ is everywhere dense and open.*

Proof. This set is open since the topology of Γ is stronger than the point-wise topology , furthermore, it is everywhere dense since there exists $g \in \Gamma$ with $g(x) \neq g(x')$; the sequence $g_n = f + \frac{1}{n} \cdot g$ separates x and x' when $f(x) = f(x')$, and converges near f .

Theorem 2.3.- *Let us assume* $F_E(Z)$ *to be locally bounded with* E *having a countable basis of open sets. Then for every natural Frechet space* Γ *in* $F_X(X,Z)$ *such that* X *is separated by* Γ *and is a maximal* Γ*-extension, there exists* $f \in \Gamma$ *such that* X *is f-maximal.*

Proof. Let (ω_n) be a countable basis of domains for E ; given $a_n \in \omega_n$ for each n , the set $Q = \bigcup p^{-1}(a_n)$ is countable by the corollary of proposition 1.2. Let (x_n) now an everywhere dense sequence in X , whose existence is given by proposition 1.2. We denote by N the set of pairs of integers (m,n) such that $p(x_m) \in \omega_n$ and $\Gamma_{\omega_n, x_m} \neq \Gamma$. With the notations of proposition 2.3, we know by proposition 2.3 and 2.4 that there exists $f \in \Gamma$ which does not belong to any Γ_{ω_n, x_m} with $(m,n) \in N$ and separates Q . We shall prove that every f-extension (X',p') of X is isomorphic to X . Let u be the morphism from X to X' and f' the extension of f to X' .

a) u is one to one :

Let x_1 and x_2 be two points of X with $u(x_1) = u(x_2) = x'$. There exist some neighbourhood Ω_1 , Ω_2 , Ω' of x_1 , x_2 , x' which are homeomorphic to some connected neighbourhood ω of the point $a = p'(x') = p(x_1) = p(x_2)$, by the projections p and p' ; furthermore we have $u(\Omega_1) = u(\Omega_2) = p_{x'}^{-1}(\omega)$, and there exists some a_n belonging to ω.

If $x_1 \neq x_2$, then $\Omega_1 \cap \Omega_2 = \emptyset$, $y_1 = p_{x_1}^{-1}(a_n)$ and $y_2 = p_{x_2}^{-1}(a_n)$ are both in Q which implies $f(y_1) \neq f(y_2)$; on the other hand $f(y_1) = f' \circ u(y_1) = f' \circ u(y_2) = f(y_2)$, since $u(y_1) = u(y_2) = p'^{-1}_{x'}(a_n)$; thus $x_1 = x_2$.

b) u is onto :

Now, we can assume X is a domain in X' and u the imbedding map.

If $X \neq X'$, there would exist x_o belonging to the boundary of X , a neighbourhood ω of x_o which is homeomorphic with some ω_n by p' , f' bounded on ω, $x_m \in \omega \cap X$. Since X is a maximal Γ-extension, $X \cup \omega$ cannot be a Γ-extension and $\Gamma_{\omega_n, x_m} \neq \Gamma$; therefore $f \notin \Gamma_{\omega_n, x_m}$, is impossible since $(f \circ p_{x_m}^{-1})_{p(x_m)} = (f' \circ p'^{-1}_{x_m})_{p(x_m)}$.

CHAPTER III : ANALYTIC MAPPINGS

§ 1.- Basic properties.

Here, some definitions and propositions are given with short proofs, we refer for them to [40,70[. The space E (resp. Z) is a Hausdorff locally convex vector space (abrev. c. v. s.) with $K = \mathbb{R}$ or $\mathbb{C}$ as the field (resp. sequentially-complete c. v. s. with $\mathbb{C}$ as the field). Let ω be an open set in E.

Properties of Gateaux analytic mappings.
E has $\mathbb{C}$ as its field.

Definition 3.1.- A mapping $f : \omega \to \mathbb{C}$ is Gateaux-analytic iff for any z' in the adjoint space Z' , any $x \in \omega$ and $a \in E$, the mapping $t \to z' \circ f(x + at)$ is holomorphic in a neighbourhood of the origin in $\mathbb{C}$.

*Proposition 3.1.- Given $a_1 ,\dots, a_n$ in E , for any fixed x , there exists c_α in Z such that for any $\pi \in N(Z)$ the series $\Sigma\, c_\alpha t^\alpha$ is summable for ~~π~~ near $f(x + t_1 a_1 + \dots + t_n a_n)$ and for t small enough ; here $t = (t_1 ,\dots, t_n)$ and α is an n-multi-index.
For $\alpha = (1, 1, \dots, 1)$, the associated coefficient is denoted by $f_n(x, a_1, a_2, \dots, a_n)$; for $n = 1$ and $\alpha = k$, the associated coefficient is denoted by $f^k(x,a)$.*

Proposition 3.2.-

a) $f^n(x, a, \dots, a) = n! \,.\, f^n(x,a)$.

b) The mapping $(a_1 ,\dots, a_n) \to f^n(x, a_1 ,\dots, a_n)$ is n-linear.

c) The mappings $x \to f^n(x,a)$ and $x \to f^n(x, a_1 ,\dots, a_n)$ are Gateaux-analytic.

d) $f^n(x,a) = \frac{1}{2\pi} \int_0^{2\pi} f(x + a\, e^{i\theta})\, e^{-ni\theta} d\theta$, for all $a \in \omega(x)$, the biggest balanced open set with x as center and contained in ω.

e) Any Gateaux-analytic mapping satisfies property (A).

f) Let π a continuous semi-norm on Z be given. If $\pi \circ f$ is continuous at $x \in \omega$, then $\pi \circ f$ is continuous in a neighbourhood of x and *each map* $a \to \pi \circ f^n(x,a)$ *is continuous on* E .

Proofs. The proposition 3.1 is a consequence of Hartogs theorem and the equivalence of weak-holomorphy and holomorphy in $\mathbb{C}^n$.

Proposition 3.2.a) Since the statement is obvious for $n = 1$, we may assume that $(n-1)!\, f^{n-1}(x+ta;a) = f^{n-1}(x+ta;\, \underbrace{a,\dots,a}_{(n-1)\text{ times}})$ whenever $x+ta \in \omega$. But $f^{n-1}(x+ta;\dots)$ in the left hand side (resp.: the right hand side) is the coefficient of u^{n-1} (resp.: $u_1 \dots u_{n-1}$) in the Taylor expansion in u (resp.: $u_1,\dots,u_{n-1}$) of $f[x + (t+u)a]$ (resp.: $f[x + (t + u_1 + \dots + u_{n-1})a]$) , which is an analytic function of (t,u) [resp.: $(t,u_1,\dots,u_{n-1})$] on some neighbourhood ot he origin in $\mathbb{C}^2$ (resp.: $\mathbb{C}^n$) . Therefore, for sufficiently small $|t|$, $f^{n-1}(x+ta;a)$ (resp.: the right hand side) is the sum of a power series in t , where the coefficient of t is the coefficient of tu^{n-1} (resp.: $tu_1 \dots tu_{n-1}$) is the expansion of $f[x + (t+u)a]$ (resp.: $f[x + (t + u_1 + \dots + u_{n-1})a]$), namely $nf_n(x;a)$ [resp.: $f^n(x;\underbrace{a,\dots,a}_{n\text{ times}})$] .

b) c) By a similar argument : for sufficiently small $|u|$, $f^{n-1}(x+ua;a_1\dots,a_{n-1})$ is the sum of a power series in u , hence c), and its first order derivative with respect to u , for $u = 0$, is $f^n(x;a,a_1,\dots,a_{n-1})$; therefore $a_1 \to f^{n-1}(x+ua;a_1,\dots,a_{n-1})$ linear entails $a_1 \to f^n(x;a,a_1,\dots,a_{n-1})$ linear and b) will follow from the linearity of $[a \to f^1(x;a)]$.

Consider the Taylor expansion, around the origin in $\mathbb{C}^2$, of $f(x + ua + u'a')$: since the coefficient of u (resp.: u') is unaltered by setting $u' = 0$ (resp.: $u = 0$) , its value is $f^1(x;a)$ [resp.: $f^1(x;a')$]; then, for constant λ , $\lambda' \in \mathbb{C}$, the coefficient of u in the expansion of $f[x + u(a + \lambda'a')]$ is $\lambda f^1(x;a) + \lambda' f^1(x;a')$.

d) is a consequence of the summability of Taylor expansion of the map $t \to f(x+ta)$, for all $|t| < 1$ and $a \in \omega(x)$.

e) is a consequence of connectedness of E by polygonal lines and property (A) coming up in finite dimensional case.

f) This is a local property and we can assume that $\pi \circ f$ is bounded in ω by M .

Let $x \in \omega$ be given, then $|\pi \circ f(x+a) . \pi \circ f(x)| \leqslant \sum_{n \geqslant 1} \pi \circ f^n(x,a) < \frac{M.\varepsilon}{1-\varepsilon}$ all $a \in \varepsilon.\omega(x)$ by d). The continuity of $a \to \pi \circ f^n(x,a)$ is a consequence of local boundness of this mapping by a) and b).

Properties of analytic mappings.

Definition.- A map $f : \omega \to Z$ is analytic iff it is continuous and Gateaux-analytic. The presheaf of such maps will be denotes by $\mathcal{O}_E(Z)$.

Proposition 3.3.- Let $f \in \mathcal{O}_E(\omega, Z)$ be given.

a) The maps $f^n(a_1,\dots,a_n) : x \to f^n(x, a_1, a_2, \dots, a_n)$ all $(a_i) \in Z^n$ and $f^n(a) : x \to f^n(x,a)$ belong to $\mathcal{O}_E(\omega, Z)$, all $a \in Z$.

b) The maps $a \to f^n(x,a)$ and $(a_1,a_2,\dots,a_n) \to f^n(x,a_1,a_2,\dots,a_n)$ are continuous.

c) Given $\pi \in N(Z)$, the serie $\Sigma f^n(x,a)$ is uniformly summable near $f(x+a)$ in any compact set contained in $\omega(x)$ and uniformly summable for a in some neighbourhood V_π of the origin in E .

d) $\mathcal{O}_E(\omega, Z)$ is closed in $C_E(\omega, Z)$ and $f \to f^n(a)$ is continuous with respect to compact open topology.

As consequence of proposition 3.2.e and 3.3.d, $\mathcal{O}_E(Z)$ satisfies property (A) and (B) .

Proof. a) Given $x \in \omega$ and $p \in N(Z)$: if B and A_n (for each integer ger $n > 0$) are balanced neighbourhood of the origin in ε such that $\underbrace{A_n + \dots + A_n}_{\textbf{n times}} \subset B \subset \omega - x$ and $p \circ f \leqslant M$ on $x + B$, then $p \circ f^n(a) < M$ on $x + A_2$ for a A_2 , $p \circ f^n(a,a_2,\dots,a_n) \leqslant M$ on $x + A_2$ for $a \in A_2$, $a_1,\dots,a_n \in A_{n+1}$.

For any given a or $a_1,\dots,a_n \in E$, by n-homogeneity or n-linearity we have $p \circ f_n(a)$ or $p \circ f^n(a,a_2,\dots,a_n)$ locally bounded on ω ; since $f^n(a)$ and $f^n(a,a_2,\dots,a_n)$ are Gateaux-analytic by proposition 3.2.c), they also belong to $\mathcal{O}(\omega,Z)$.

b) is a consequence of proposition 3.2.f).

c) Let K be a compact set in ω and Δ the unit disc in $\mathbb{C}$. All $a \in K$, 1 and a have neighbourhoods T in $\mathbb{C}$ and A in E such that $\omega(x) \supset TA$; if a finite union of sets A_i contains K and the corresponding neighbourhoods T_i of 1 contain the disc then $\omega(x)$ contains $(1+\alpha.\Delta).K$, hence contains the compact set $(1+\alpha)\Delta.K$ since $\omega(x)$ is balanced.

Given $p \in N(Z)$; $p \circ f \leqslant M$ on $x + (1+\alpha)\Delta.K$ implies $p \circ f^n(x;a) \leqslant M$ $\forall a \in (1+\alpha)\Delta.K$, hence $p \circ f^n(x;a) \leqslant M/(1+\alpha)^n$ $\forall a \in K$, $n \in \mathbb{N}$, and the uniform summability.

d) Given a compact set $K \subset \omega$ and $p \in N(Z)$: if a is chosen sufficiently near the origin in E , $\{x+ta : x \in K, t \in \Delta\} = \tilde{K}$ is a compact subset of ω , and :

$$\sup_{x \in K} p \circ f^n(x,a) \leqslant \sup_{x \in \tilde{K}} p \circ f(x)$$, by proposition 3.2.d) .

The closeness of $\mathcal{O}(\omega,Z)$ in $C(\omega,Z)$ is an obvious consequence of the same property in finite dimensional case.

Proposition 3.4.- Any Gateaux-analytic mapping which is locally bounded is continuous. The presheaf of locally bounded analytic mappings will be denoted as $\mathcal{O}^b_E(Z)$. Indeed, $\mathcal{O}^b_E(Z) = \mathcal{O}_E(Z)$, whenever E is normed. By proposition 3.3.d $\mathcal{O}^b_E$ satisfies property (B) .

Proof. Let U be a balanced neighbourhood of $x \in \omega$ where f is bounded. For any $\pi \in N(Z)$ we have $\pi(f(x+A) - f(x)) \leqslant \sum_{n \geqslant 1} \pi \circ f^n(x,a)$, all $a \in U$. Then the right hand is less than $\varepsilon > 0$ for a belonging to $\lambda.U$ with a suitable $\lambda > 0$.

Proposition 3.5.- [66] Let f be a Gateaux-analytic map from ω to Z be given.

a) If the maps $a \to f^n(x,a)$ are continuous at some $x \in \omega$ for all $n \in \mathbb{N}$, if E is a Baire space then f is continuous at x .

b) If f is weakly analytic (that is $\zeta \circ f \in \mathcal{O}_E(\omega,\mathbb{C})$ all $\zeta \in Z'$) and E is metrizable then f belongs to $\mathcal{O}_E(\omega,Z)$.

Proof. a) Let $p \in N(Z)$ be given ; $\forall a \in \omega(x)$, since the expansion $\sum_n f^n(x;a)$ is summable, $g(a) = \sup_{n \geqslant 0} p \circ f^n(x;a)$ is finite.

By the assumption, each function $a \to p \circ f^n(x;a)$ is continuous on E , therefore g is l.s.c. and finite on $\omega(x)$; since $\omega(x)$ is a Baire space, there exist $a_o \in \omega(x)$, a balanced neighbourhood B of the origin in E and a number M , such that $g \leqslant M$ on $a_o + B$, or $p \circ f^n(x;a) \leqslant M$ $\forall a \in a_o + B$, $n \in \mathbb{N}$; we shall get the same inequality $\forall a \in B$, and the proof will be over.

Since $[u \to f^n(x;a + ua_o)] \in \mathcal{O}(\mathbb{C},Z)$, $u \to p \circ f^n(x;a + ua_o)$ is subharmonic on $\mathbb{C}$, hence $p \circ f^n(x;a) \leqslant$ mean value of $p \circ f^n(x;a+ua_o)$ for $|u| = 1$. But, by n-homogeneity, for $|u| = 1$: $p \circ f^n(x;a+ua_o) = p \circ f^n(x;a_o + \frac{a}{u}) \leqslant M$ if $a \in B$.

b) The property is known when E is finite dimensional, then f is Gateaux-analytic. Then weak continuity of f entails that f(K) is bounded for any compact set. Let $\pi \in N(Z)$ be given, since E is metrizable $\pi \circ f$ is locally bounded and we can apply the proof of proposition 3.2.f).

Remark.- Without assumption concerning E, a) and b) in proposition 3.5 are false.

a)- We take for E the space $C_{o,o}$ of sequence (x_n) in $\mathbb{C}$ which vanishes for n large enough with the uniform norm and $Z = \mathbb{C}$. The function $f(x) = \sum_{n>1} (n\, x_n)^n$ is Gateaux-analytic on E and $f^n(o,x) = (nx_n)^n$. Nevertheless $f(\frac{1}{n}\ell_n) = 1$ and $\frac{1}{n}\ell_n$ converges to 0 with (ℓ_n) as the canonical basis of $C_{o,o}$.

b)- We take $E = Z_\sigma$, that is Z equipped with the weak star topology, and f = identity.

Properties of real analytic mappings.

E has $\mathbb{R}$ as its field.

Given an open set ω in E and a map $f : \omega \to \mathbb{C}$ such that for any $x \in \omega$ there exists a sequence of continuous, homogeneous polynomials $a \to f^n(x,a)$, with degree = n, and the series $\sum_{n\geqslant 0} f^n(x,a)$ is uniformly convergent to f(x+a), all a in a suitable convex, balanced, neighbourhood U of 0 E. We define $f^n(x,a,b^k) = f^n(x,a(n-k)\text{times}, b(k\text{-times}))$ and $\tilde{f}^n(a+ib) = \sum [\binom{n}{4p} . f^n(a,b^{4p}) - \binom{n}{4p+2} f^n(a,b^{4p+2}) + i \binom{n}{4p+1} . f^n(a,b^{4p+1}) - i \binom{n}{4p+3} f^n(a,b^{4p+3})]$. The previous map $\tilde{f}^n$ is a continuous, homogeneous polynomial on the complexified space $\tilde{E}$ of E and the series $\sum_{n\geqslant 0} \tilde{f}^n(x,a+ib)$ is uniformly convergent for a and b in 1/2.e U because the elementary inegality [32] : $\sup |f^n(a_1,\dots,a_n)|$, $a_i \in U \leqslant \frac{n^n}{n!} \sup |f^n(a)|$, $a \in U$.

Therefore the series $\sum_{n\geqslant 0} \tilde{f}^n(x,a+ib)$ has a sum which belongs to $\mathcal{O}_{\tilde{E}}(\mathbb{C})$, and the following definition is convenient.

Definition 3.2.- A mapping $f : \omega \to Z$ is analytic on the open set ω, if there is a neighbourhood $\tilde{\omega}$ of ω in the complexified space $\tilde{E}$ of E

and $f \in \mathcal{O}_E(\tilde{\omega}, Z)$ *such that* f *is the restriction of* $\tilde{f}$ *to* ω .
The presheaf of real analytic mappings will be denoted by $A_E(Z)$.

Proposition 3.6.- The propositions 3.2.a-b-e and 3.3.b (c, second part) are satisfied by $A_E(Z)$ *instead of* $\mathcal{O}_E(Z)$.
This is an obvious consequence of the definition.

Remark.- It is well known that $A_E(Z)$ does not satisfy property (B). Nevertheless, some sub-presheaf could be satisfy it ; for instance, the space of harmonic functions satisfied (B) , $(E = \mathbb{R}^n$ and $Z = \mathbb{C}$).

§ 2.- Some remarks about analytic extensions.

Let us assume $E = \mathbb{C} \times F$ with F a real c. v. s. ; for any open set ω , let denote by $\Gamma(\omega)$ the set of functions $f(z,x)$ which belong to $A_E(\omega, \mathbb{C})$ and are such that $z \to f(z,x)$ is holomorphic on the sections ω_x , $x \in F$.

Theorem 3.7.- Let us assume F *to be infinite dimensional. Given a compact set* K *in* E *such that* K_x *would be compact in* ω_x *, all* $x \in F$ *, then* $\Gamma(\omega)$ *is an extension of* $\Gamma(\omega - K)$.

Proof. Let $f \in \Gamma(\omega - K)$ be given.

1) Firstly, a maximal open set Ω is picked for the relation :
$\Omega_1 < \Omega_2$ whenever $\omega - K \subset \Omega_1 \subset \Omega_2 \subset \omega$ and there exist $g_2 \in \Gamma(\Omega_2)$, $g_1 \in \Gamma(\Omega_2)$ with $g_2 = g_1 = f$ on $\omega - K$ and $g_2 = g_1$ on Ω_1 .

Now, given $g \in \Gamma(\Omega)$, with $g = f$ on $\omega - K$ and $(z_0, x_0) \in \omega - \Omega$, when this set is not empty. Since K_{x_0} is a compact set in ω_{x_0} , there exist neighbourhoods Δ of $0 \in \mathbb{C}$ and V of $0 \in F$ such that $K_{x_0} + 2\Delta$ is contained in ω_x , all $x \in x_0 + V$. Furthermore V can be chosen such that K_x is contained in $K_{x_0} + \Delta$ for all $x \in V$. Then we have :

$$K_x \subset K_{x_0} + \Delta \subset K_{x_0} + 2\Delta \subset \omega_x \text{ , all } x \in V .$$

Now, a function $\varphi(z)$ with value 1 on $K_{x_0} + \Delta$, with support in $K_{x_0} + 2\Delta$, indefinitely differentiable, is picked out ; let $v(z,x)$ be defined by :

$$(1) \quad v(z,x) = \frac{1}{2\pi i} \int g(\tau,x) \cdot \frac{\partial}{\partial\bar{\tau}} \varphi(\tau) \cdot \frac{1}{\tau - z} \cdot d\tau \cdot d\bar{\tau} \text{ , all } x \in V .$$

On the other hand, let K_F be the canonical projection of K on F ,

(1) entails :

$$(2)\quad v(z,x) = \frac{1}{2\pi i}\int \frac{\partial}{\partial\bar\tau}\ (g(\tau,x)\ .\ \frac{\partial}{\partial\bar\tau}\varphi(\tau)).\ \frac{1}{\tau-z}\ .\ d\tau\ .\ d\bar\tau = g(z,x).\varphi(z)$$

all $(z,x) \in \mathbb{C} \times (V - K_F)$.

Let $h(z,x)$ be defined by $h(z,x) = (1 - \varphi(z))\ g(z,x) + v(z,x)$.

This h is continuous in $(\mathbb{C} \times V) \cap \omega$ and $h = g$ in $(\mathbb{C} \times (V - K_F)) \cap \Omega$ by (2) ; since K_F is compact and so has no interior, we have $h = g$ in $(\mathbb{C} \times V) \cap \Omega$. Since, Ω has been chosen maximal, then the proof is complete after checking the analyticity of h at (z_o, x_o) .

In $(z_o + \Delta) \times V$, the following relation is coming up :

$$h = v = \frac{1}{2\pi i}\int_{(K_{x_o} + 2\Delta) - (K_{x_o} + \Delta)} g(\tau,x)\ .\ \frac{\partial}{\partial\bar\tau}\ \varphi(\tau)\ .\ \frac{1}{z-\tau}\ .\ d\tau\ .\ d\bar\tau\ .$$

Noting the analyticity of $(\tau,z,x) \to g(\tau,x)\ \frac{1}{z-\tau}$ in $\left[(K_{x_o} + 2\Delta) - (k_{x_o} + \Delta)\right] \times (z_o + \Delta) \times V$, then v is also analytic in $(z_o + \Delta) \times V$.

<u>Comment</u>.- If F is finite dimensional, the previous result can be false. For instance, $\frac{1}{z-x}$ for $|z| < 1$ and $|x| < 1$. Some more requirements are necessary about K in the finite dimensional case, [cf. 8] .

The next result shows how a new fact again arises in the infinite dimensional case. In $E = \mathbb{R}^n$, any open set is a maximal extension for $A_E(\mathbb{C})$, but it is not true in $\mathbb{R}^{\mathbb{N}}$. We shall denote by π_n the canonical mapping from $\mathbb{R}^{\mathbb{N}}$ into $\mathbb{R}^n$.

<u>*Theorem*</u> *3.8.- [45,75] . Given a domain ω in $\mathbb{R}^{\mathbb{N}} = E$ and $f \in A_E(\omega, \mathbb{C})$, there exists an integer p such that each $x \in \omega$ has a neighbourhood V with $g \in A_{\mathbb{R}^p}(\pi_p(V), \mathbb{C})$ with satisfies $f = g \circ \pi_p$ on V .*

<u>Proof</u>. Firstly, a local statement will be established ; therefore, we can assume f belonging to $\mathcal{O}_{\mathbb{C}^{\mathbb{N}}}(\mathbb{C})$. There exists a neighbourhood V of 0 in some $\mathbb{C}^p$, such that f is bounded in $\pi^{-1}(V)$. For any $a \in \pi^{-1}(V)$ and $z \in \mathbb{C}$, the function $z \to f[\pi_p(a) + z(a - \pi_p(a)]$ is entire and bounded ; so it is constant. Therefore $f \circ \pi_p(a) = f(a)$.

Now, denote by $n(x)$ the smallest integer p for which the local statement is true at x ; $n(x)$ is locally constant and the proof is complete.

Corollary 3.8.- Let I *a set and* $f \in \mathcal{O}(\mathbb{C}^I)$ *be given. For all finite subset* Λ *of* I*, all neighbourhood* V *of the origin in* $\mathbb{C}^\Lambda$ *such that* f *is bounded in* $V \times \mathbb{C}^{I-\Lambda}$*, we have* $f(x) = f[\pi_\Lambda(x)]$*, all* $x \in \mathbb{C}^I$*,* π_Λ *is the canonical projection from* $\mathbb{C}^I$ *onto* $\mathbb{C}^\Lambda$.

Proof. The first part of the proof in the previous theorem is true for a general set I. Therefore we have $f(x) = f[\pi_\Lambda(x)]$ for all $x \in V \times \mathbb{C}^{I-\Lambda}$. The both hands belongs to $\mathcal{O}(\mathbb{C}^I)$ then the equality is true for all x.

Theorem 3.9.- [45]. For any open set ω *and any affine, closed, infinite codimensional subspace* L *in* $\mathbb{R}^{\mathbb{N}}$*,* ω *is an extension of* $\omega - L$ *for* $A_{\mathbb{R}^{\mathbb{N}}}(\mathbb{C})$.

Proof. Firstly, ω can be assumed connected and $\omega - L$ also connected. Actually, given x_1 and x_2 in $\omega - L$, we can join them by a polygonal line $[x_1, a_1, \ldots, a_n, x_2]$ in ω. The affine space $[x_1, x_2] \oplus L$ does not fill $\mathbb{R}^{\mathbb{N}}$, so there is a_1' near a_1 such that $[x_1, a_1']$ is contained in ω and not contained in $[x_1, a_1] \oplus L$; therefore $[x_1, a_1']$ does not cut L. After n steps, we have constructed a polygonal line $[x_1, a_1', \ldots, a_{n+1}']$ in $\omega - L$, with a_{n+1}' near x_2.

Now, we take the integer p given by theorem 3.8 applied for $\omega - L$. Using a Zorn argument as in theorem 3.7, given $x_0 \in \omega \cap L$, we must construct a neighbourhood V of x_0 and $g \in A_{\mathbb{R}^{\mathbb{N}}}(\omega, \mathbb{C})$ with $g = f$ in $V \cap (\omega - L)$; f being given in $A_{\mathbb{R}^{\mathbb{N}}}(\omega - L, \mathbb{C})$.

There exists an integer $n \geqslant p$ and a neighbourhood W of $\pi_n(x_0)$ in $\mathbb{R}^n$ with $V = \pi_n^{-1}(W)$ contained in ω.

By theorem 3.8, there exists $m \geqslant n$, a neighbourhood V' of $\pi_m(x_1) = \pi_n(x_0)$ in $\mathbb{R}^m$, $g \in A_{\mathbb{R}^p}[\pi_p(V'), \mathbb{C}]$ such that $f = g \circ \pi_p$ in $\pi_m^{-1}(V')$.

Furthermore, W can be chosen such that $\pi_p(\omega)$ is contained in $\pi_p(V')$, then $g \circ \pi_p$ belongs to $A_{\mathbb{R}^{\mathbb{N}}}(V, \mathbb{C})$. Noting that $V \cap (\omega - L) = V - L$, $f = g \circ \pi_p$ in $V - L$ since $V - L$ is connected and $f = g \circ \pi_p$ near x_1 which belongs to $V - L$. The proof is complete.

Before extensively studying analytic extension in the next chapter, we point out the following fact. Given a set Γ in $\mathcal{O}^b_E(X,Z)$, each Γ-extension $u : X \to X'$ for $\mathcal{O}^b_E(Z)$ is a Γ-extension for $\mathcal{O}_E(Z)$, but the converse is generally false.

Example.- Z is the Frechet space $\mathbb{C}^{\mathbb{N}}$ equipped with the product topology, E is the space of sequences $x = (x_n)$ which converge to zero, equipped with the uniform norm.

Let f_p be defined in $\mathcal{O}(E, \mathbb{C})$ by $f_p(x) = \sum_{n>0} (x_o^p, x_n)^n$ for each integer p , the sequence $f = (f_p)$ is in $\mathcal{O}(E, \mathbb{C}^{\mathbb{N}})$, bounded in the unit ball B , but not bounded in any neighbourhood of each x which $\|x\| > 1$. Setting $\Gamma = \{f\}$; $B \to E$ is a Γ-extension for $\mathcal{O}_E(Z)$, but not for $\mathcal{O}^b_E(Z)$.

CHAPTER IV : FRECHET SPACES OF COMPLEX ANALYTIC MAPPINGS

Z is a complex Frechet c. v. s. and X is a manifold spread over a complex c. v. s. E .

When E is finite dimensional, $\mathcal{O}_X(X,Z)$ is a class of type $A_{\mathcal{U}}$ with $\mathcal{U}$ a countable covering of X by relatively compact sets and so is a natural Frechet space for the compact open topology. When E is infinitely dimensional, it is another matter. In a Banach space E such that every sequence in the adjoint space E' contains a pointwise convergent subsequence, the bounding sets for $\mathcal{O}_E(E, \mathbb{C})$ are relatively compact [24] ; hence, $\mathcal{O}_E(E, \mathbb{C})$ is never uniformly bounded and so is never a natural Frechet space by theorem 2.1. Therefore, natural Frechet spaces are generally thin in $\mathcal{O}_X(X,Z)$; nevertheless their properties are nice enough to be described now.

§ 1.- Strong invariance by derivation.

Definition 4.1.- A linear topological vector space Γ in $\mathcal{O}_X(X,Z)$ will be called strongly invariant by derivation (Abbrv. s. i. d.) if :

- $f^n(a) \in \Gamma$, all $f \in \Gamma$, $a \in E$

- The mappings $(a,f) \to f^n(a)$ from $E \times \Gamma$ into Γ are equicontinuous when n describes the integers.

The next proposition is established to justify the above definition.

Proposition 4.1.- Let Γ be a natural Frechet space in $\mathcal{O}_X(X,Z)$ which is invariant by derivation ; then

- the mappings $f \to f^n(a)$ from Γ into Γ are continuous for any n and $a \in E$;

- if E is a Frechet c. v. s, the mappings $(a,f) \to f^n(a)$ are continuous for all n from $E \times \Gamma$ into Γ .

Proof.- By the polarization identity between $f^n(a)$ and $f^n(a_1,a_2,\ldots,a_n)$, this last mapping belongs to Γ . We can apply the closed graph theorem to prove that the (n+1) linear mapping $(a_1,\ldots,a_n;f) \to f^n(a_1,a_2,\ldots,a_n)$ is separately continuous from E or Γ into Γ .

Further, since E is metrizable and Γ is a Baire space, the separate continuity implies continuity, [77] .

Now, to prove the closed graph property, taking sequence (a_k) and (f_k) which converge to a in E and f in Γ , let us suppose that $f_k^n(a)$ converges near g and $f^n(a_k)$ near h in Γ .

Since the Γ-topology is stronger than pointwise convergence and the mapping $a \to f^n(x,a)$ is continuous from E into Z , we have $h = f^n(a)$.

We know, by proposition 3.3.d, that the mapping $f \to f^n(a)$ is continuous for compact convergence, further that the Γ-topology is stronger than compact topology by corollary 2.1 ; then we have $g = f^n(a)$.

Hence, for the usual space E , the previous definition is convenient and introduces a stronger property for natural Frechet spaces. The following example shows that there are natural Frechet spaces which are invariant but not strongly invariant by derivation.

Example : Given a domain ω in $\mathbb{C}$, and let Γ be the Frechet space of holomorphic functions in ω such that all derivatives of any $f \in \Gamma$ are bounded on ω : Γ is equipped with the semi-norms $P_n(f) = \|d^n/dz^n . f\|_\omega$.

If the boundary of ω has a singular point for simultaneous continuation of Γ , then Γ is not strongly invariant. Actually, suppose the converse ; we should have :

Given $\varepsilon > 0$, there exist N and $\eta > 0$ such that :

$$\|f^n(a)\| = \frac{1}{n!} \left\|\frac{d^n}{dz^n} f\right\| |a|^n < \varepsilon$$

for all $|a| < \eta$, all $f \in \Gamma$ with $\sum_{k<N} \left\|\frac{d^k}{dz^k} f\right\|_\omega < \eta$, all n .

Then, every $f \in \Gamma$ has a Taylor expension at any $x \in \omega$ with $R > \eta$ as radius of convergence. This is impossible near a singular point of the boundary of ω.

We can take for ω , the half plane $\operatorname{Re} z < 0$, where the origin is a singular point for $\exp \frac{1}{z}$ which belongs to Γ .

We study the existence of some type of Frechet spaces Γ in $\mathcal{O}_X(X,\mathbb{C})$. The adjoint space E' of E equipped with strongly topology (resp. topology induced by that of Γ) is denoted by E'_β (resp. E'_Γ).

We say Γ contains E' if $\xi \circ p$ belongs to Γ , all $\xi \in E'$.

Proposition 4.2.- If there exists a natural, invariant by derivation, Frechet space Γ in $\mathcal{O}_X(X,\mathbb{C})$ such that Γ contains E', then :

a) there exists a countable, fundamental, system of bounded sets in E.

b) E'_Γ is a Frechet space finer than E'_β.

Proof. First we prove that E'_Γ is a Frechet space. A Cauchy sequence ξ_n in E'_Γ is such that $\xi_n \circ p$ converges to some f in Γ. By proposition 4.1.a), $(\xi_n \circ p)'(a)$ is convergent to $f'(x,a)$, all $a \in E$; that is $\xi_n(a)$ is convergent to $f'(x,a)$, all $x \in X$, all $a \in E$ since the topology of Γ is natural. Thus f belongs to E'. Let (V_n) be a countable basis of neighbourhoods of 0 in E'_Γ ; given B a bounded set in E, the polar set B^o is closed in E'_Γ since the topology of E'_Γ is natural ; then B^o is a barrel in E'_Γ and so a neighbourhood of 0. We have proved b).

On the other hand, there exists some V_n contained in B^o, that is B is contained in V_n^o. Noting the duality (E'_Γ, E) because of natural topology of Γ, V_n^o is weakly bounded and also bounded.

Corollary 4.2.- If E is infrabarrelled and there exists a natural, Frechet space Γ which contains E' in $\mathcal{O}_X(X,\mathbb{C})$, then E is a $\mathcal{D}.\mathcal{F}$ space and $E'_\beta = E'_\Gamma$; furthermore, among the metrizable spaces, only the normable spaces have the above properties.

Proof. It is known [51,77] that an infrabarrelled space with the property a) of proposition 4.2 is a $\mathcal{D}.\mathcal{F}$ space then E'_β is a Frechet space. Lastly since E'_Γ is a Frechet space finer than E'_β, we have $E'_\Gamma = E'_\beta$. Now, if E is metrizable, then E'_β is metrizable iff E is normable [51].

When E is normed, there exist spaces Γ with properties of the corollary. Actually, we take for Γ a space of type $A_{\mathcal{U}}$, with $\mathcal{U}$ an admissible, countable covering of X, such that $p(\omega)$ is bounded, all $\omega \in \mathcal{U}$.

§ 2.- A general type of strongly-invariant by derivation spaces.

In the following, the notation below will be used :

Given a set T in X and V a balanced neighbourhood of the origin in E, we write $T + V \subset X$ iff for every $x \in T$, p is a homeomorphism of a neighbourhood of X onto $p(x) + V$, and $T + V = \{p_x^{-1}[p(x)+V],\ x \in T\}$;

we write also $T + Q$ with Q a part of E , when Q is contained in some previous V .

Definition 4.2.- *A covering $\mathcal{U}$ of X by open sets will be called admissible if, for any $\omega \in \mathcal{U}$ there exists V balanced neighbourhood of the origin in E such that $\omega + V$ is contained in some $\omega' \in \mathcal{U}$.*

Proposition 4.3.- *Whenever $\mathcal{U}$ is an admissible covering of X , the associated class $A_{\mathcal{U}}$ is strongly invariant by derivation.*

Proof. Given $q \in N(Z)$ and $\omega \in \mathcal{U}$; we have, by proposition 3.2. d)

$$\sup_{x \in \omega} q \circ f^n(x,a) \leqslant \frac{1}{2\pi} \int_0^{2\pi} q \circ f \circ p_x^{-1} \left[p(x) + a \cdot e^{i\theta}\right] \cdot e^{-ni\theta} \, d\theta$$

$$\leqslant \sup_{x \in \omega'} q \circ f(x), \text{ for all } a \in V .$$

Here, ω' and V are the sets associated with ω by the previous definition.

Proposition 4.4.- *Given a covering $\mathcal{U}$ of X by open sets and a basis $\mathcal{V}$ of balanced neighbourhoods of the origin in E ; there exists another covering $\mathcal{U}'$ of X which is admissible and finer than $\mathcal{U}$.*

If $\mathcal{U}$ is countable, $\mathcal{U}'$ can be chosen countable.

Proof. Given $V \in \mathcal{V}$ and $\omega \in \mathcal{U}$, we denote $\omega(V) = \{x \in \omega \mid \exists\, r(x) > 1$ with $x + r(x).V \subset \omega\}$.

It is easy to verify that $\omega(V)$ is open and $\omega(V) + 1/2.V$ is included in $\omega(1/2.V)$; then we have constructed an admissible covering of X by $\omega(V)$ when V describes $\mathcal{V}$.

Corollary 4.4.- *If Z is Banach space and E is metrizable, any natural Frechet space in $\mathcal{O}_X(X,Z)$ can be continuously imbedded into a natural s. i. d. Frechet space, that is a class of type $A_{\mathcal{U}}$ with $\mathcal{U}$ an admissible and countable covering of X .*

Proof. It is an obvious consequence of theorem 2.2, propositions 4.2 and 4.3.

Proposition 4.5.- *Given a bounded set B in $\mathcal{O}_X(X,Z)$ equipped with compact topology, and assume Z is a Banach space and E is metrizable ; then B is contained into a convenient class of type $A_{\mathcal{U}}$ with $\mathcal{U}$ as an admissible and countable covering of X .*

Proof. Since X has countable basis of neighbourhoods of each point, and Z is complete, $C_X(X,Z)$ is complete for the compact open topology, and $\mathcal{O}_X(X,Z)$ is also complete by proposition 3.3.d). Therefore the space $E_{\tilde{B}}$, which is spanned by the closed convex hull $\tilde{B}$ of B, is a Banach space for the Minkowski-norm associated with $\tilde{B}$, and its topology is stronger than compact topology ; hence, $E_{\tilde{B}}$ is a particular natural Frechet space. Now, since Z is a Banach space, we can apply the previous corollary.

Remark. By the previous corollary, a bounded set in $\mathcal{O}_X(X,Z)$ for the compact open topology is locally uniformly bounded when Z is normed. Then, we can ask the same property for a bounded set in $\mathcal{O}^b_X(X,Z)$ when Z is not normed. But it is not true as the following example will show.

An example : Z is the product space $\mathbb{C}^{\mathbb{N}}$; E is the space of sequences which converge to 0 with the norm of ℓ_∞.

For each pair of integers (k,p), let us define $f_{k,p}$ by

$$f_{k,p}(x) = \sum_{n>0} \left(\frac{k.p}{\sqrt{k^2+p^2}}\, x_n\right)^n \text{, which belongs to } \mathcal{O}(E,\mathbb{C}) .$$

The sequence $p \to f_{k,p}$ defines a mapping F_k which belongs to $\mathcal{O}(E,Z)$.

- The sequence F_k is bounded on every compact subset K of E.

For each p there exists N_p such that $|a_n| < \frac{1}{2p}$ for all $n > N_p$ and every $a \in K$; denote by M_p the upperbound of $\left|\sum_{n \leq N_p} p(a_n)^n\right|$ when a describes K.

Then we have : $|f_{k,p}(a)| \leq M_p + \frac{1}{2^{N_p}}$ for all k and all $a \in K$.

- Each F_k is locally bounded.

Given $a = (a_n) \in E$ and $\varepsilon = (\varepsilon_n) \in E$ with $\|\varepsilon\| < \frac{1}{4k}$; there exists N such that $|a_n| < \frac{1}{4k}$ for $n > N$.

Then, we have :

$$|f_{k,p}(a+\varepsilon)| \leq \sum_{n>0} k^n(|a_n| + \tfrac{1}{4k})^n \leq \sum_{n\leq N} k^n(|a_n| + \tfrac{1}{4k})^n + \frac{1}{(2k)^N} .$$

- Nevertheless, the sequence F_k is not uniformly bounded around the origin.

$$\sup_{\|a\|<\varepsilon} |f_{k,p}(a)| = \sum_{n\geqslant 0} \left(\frac{k.p}{\sqrt{k^2+p^2}}\varepsilon\right)^n ; \quad \sup_k \, \sup_{\|a\|<\varepsilon} |f_{k,p}(a)| =$$

$$\sum_{n>0} (p\varepsilon)^n = +\infty \quad , \text{ for } \varepsilon\cdot p > 1 .$$

§ 3.- Analytic extensions.

Given a Frechet space Γ in $\mathcal{O}_X(X,Z)$, and $u : X \to X'$ and Γ-extension for $\mathcal{O}_E(Z)$. Any $\pi \in N(\Gamma)$ defines a semi-norm on $u^*(\Gamma)$ by $u^* \circ \pi(f') = \pi(f' \circ u)$, all $f' \in u^*(\Gamma)$. Thus, $u^*(\Gamma)$ is a Frechet space isomorphic with Γ that we call the extended Frechet space.
However, we do not know, for the general case, if $u^*(\Gamma)$ is natural with Γ together. Nevertheless, we have this property when Γ is strongly invariant by derivation. To begin with, we point out that $u^*(\Gamma)$ is natural whenever Γ is a Frechet space $\mathcal{O}_X(X,Z)$, with Z a Frechet space and X spread over a finite dimensional space E . Actually, let j be the canonical injection $u^*(\Gamma) \to \mathcal{O}_{X'}(X',Z)_c$ and k the restriction mapping $\mathcal{O}_{X'}(X',Z)_c \to \mathcal{O}_X(X,Z)_c$; then k and $j \circ k$ are continuous together by the closed graph theorem, j is also continuous.

Proposition 4.6.- *Whenever Γ is strongly invariant by derivation, so is $u^*(\Gamma)$.*

Proof. By proposition 3.2.d, for $a \in E$ small enough, we have : $g^n(a) \circ u =$
$$\frac{1}{2\pi i}\int g \circ p^{-1}_{u(x)} \left[p \circ u(x) + a\, e^{i\theta}\right] e^{-ni\theta}\, d\theta =$$

$$\frac{1}{2\ i}\int g \circ p^{-1}_{u(x)} \left[p(x) + a\, e^{i\theta}\right] e^{-ni\theta}\, d\theta = (g \circ u)^n(x,a) \ , \quad \text{all } g \in u\,(\Gamma).$$

That is : $g^n(a) \circ u = (g \circ u)^n(a)$. On the other hand, there exists $\pi_1 \in N(\Gamma)$ and a neighbourhood V of $0 \in E$ such that : $\pi_1(g \circ u) \leqslant \eta$ and $a \in V$ entails $[(g \circ u)^n(a)] < \varepsilon$, all n . Then $u^* \circ \pi_1(g) \leqslant \eta$ and $a \in V$ entails $u^* \circ \pi[g^n(a)] < \varepsilon$, all n ; so, the proof is complete.

Theorem 4.6.- Whenever Γ is a strongly invariant by derivation, natural Frechet space, so is $u^*(\Gamma)$.

Proof. Let W be defined by $W = \{x' \in X' \mid$ the evaluation mapping $\hat{x}'$ at x' is continuous in $u^*(\Gamma)\}$.

W <u>is open</u> : given $q \in N(Z)$ and $x' \in W$, there exists $\pi_1 \in N(\Gamma)$ with $q[g(x')] \leqslant \pi_1(g \circ u)$ for any $g \in u^*(\Gamma)$; now, there exists $\pi_2 \in N(\Gamma)$ and a balanced neighbourhood V of the origin of E such that $\pi_1[f^n(a)] \leqslant \pi_2(f)$ for any $a \in V$ and $f \in \Gamma$.

We choose V small enough such that $x' + V$ will be included in X'. Then for any $a \in 1/2\, V$, we have :

$$q[g(x'+a)] = \lim_{N\to\infty} q[\sum_{n\leqslant N} g^n(x',a)] \leqslant \sum_{n\geqslant 0} \frac{1}{2^n} q[g^n(x',2a)]$$

$$\leqslant 2\pi_2(g \circ u) = 2\, u^* \circ \pi_2(g) .$$

Then, $x' + 1/2\, V$ is included in W.

Now, given $x_o \in u(x)$ and $x' \in X'$, they are joined together by a polygonal line and so, there exists a connected submanifold Y which contains x_o and x' and is spread over a finite dimensional space. We must prove that $W \cap Y$ is closed since clearly x_o belongs to W.

Given a sequence x'_n in $W \cap Y$ which converges to x'' in Y, the set $\{g \in u^*(\Gamma) \mid q[g(x'_n)] \leqslant 1$, all $n\}$ is a barrel because of boundedness of each g on the relatively compact (x'_n) and continuity of $\hat{x}'_n$; therefore this set is a neighbourhood of 0 in Γ on which we have $q \circ \hat{x}'' \leqslant 1$; then the mapping $q \circ \hat{x}''$ is continuous.

<u>*Corollary*</u> *4.6.- Let* Γ, *a s. i. d., natural, Frechet space be given in* $\mathcal{O}^b_X(X,Z)$, *if* E *is metrizable then every* Γ*-extension for* $\mathcal{O}^b_E(Z)$ *is uniformly bounded.*

<u>Proof</u>.- It is an obvious consequence of theorems 4.6 and 2.1.

<u>*Theorem*</u> *4.7.- We still asumme that* Z *is a Banach space and* E *is metrizable.*

a) Let Γ *be a natural Frechet space in* $\mathcal{O}_X(X,Z)$ *and* $u : X \to X'$ *a* Γ*-extension for* $\mathcal{O}_E(Z)$, *then the extended Frechet space* $u^*(\Gamma)$ *can be continuously imbedded in a class* $A_{\mathcal{U}'}$, *with* $\mathcal{V}'$ *an admissible and countable covering of* X'. *Thus,* $u^*(\Gamma)$ *is also natural.*

b) [79] If Γ *moreover is a class of type* $A_{\mathcal{U}}$, *with* $\mathcal{U}$ *an admissible and countable covering of* X, $\mathcal{V}'$ *can be chosen such that* $A_{\mathcal{V}'} = u^*(\Gamma)$ *as topological spaces.*

Proof.- The first part is an obvious consequence of the next sequence of previous results : Theorem 2.2, proposition 4.4, theorem 4.6, theorem 2.2, proposition 4.4.

On the other hand for the second part the range of $\mathcal{U}$ by the morphism u is an admissible and countable covering of u(X) and any f which belongs to $A_{\mathcal{U}}$ has an extension which is bounded on any set of $u(\mathcal{U})$; therefore $\mathcal{U}' = \mathcal{V}' \cup u(\mathcal{U})$ is an admissible and countable covering of X' such that the extension of $A_{\mathcal{U}}$ will be included in $A_{\mathcal{U}'}$. Further, any $f' \in A_{\mathcal{U}'}$ is the extension of $f' \circ u$ which belongs to $A_{\mathcal{U}}$

Finally, the Frechet space $A_{\mathcal{U}'}$ and the extended Frechet space of $A_{\mathcal{U}}$ have the same topology ; actually, the topology of $A_{\mathcal{U}'}$ is stronger than the topology induced by that of $A_{\mathcal{U}}$ and by applying the open mapping theorem, the proof is complete.

§ 4.- A characterisation for maximal extensions of natural strongly invariant by derivation, Frechet spaces.

Here, Z is Banach space normed by $\|\cdot\|$ and E is a metrizable c. v. s. For any open, balanced, neighbourhood V of $0 \in E$, and any set T in X , the following "boundary" functions are defined :

$$d_X^V(x) = \sup \{r \geqslant 0 \mid x + rV \text{ is contained in } X\}$$

$$d_X^V(T) = \inf d_X^V(x) \quad , \quad x \in T .$$

Definition 4.3.- Let Γ be a set in $\mathcal{O}_X(X,Z)$; the Γ-hull of a bounding set T for Γ is defined by :

$$\hat{T}(\Gamma) = \{x \in X \mid \|f(x)\| \leqslant \|f\|_T \ , \text{ all } f \in \Gamma\} .$$

Theorem 4.8.- Let Γ be a s. i. d, natural Frechet space in $\mathcal{O}_X(X,Z)$. Then any bounding set T in the Γ-maximal extension $\tilde{X}$ for the extended space $\tilde{\Gamma}$ has the following properties :

a) For some neighbourhood V of $0 \in E$, $T + V$ is a bounding set for $\tilde{\Gamma}$, contained in $\tilde{X}$.

b) For any balanced, open, neighbourhood V of $0 \in E$, we get : $d_{\tilde{X}}^V[\hat{T}(\tilde{\Gamma})] \geqslant \sup \{r > 0 \mid T + r . \Delta . a$ is a bounding set contained in $\tilde{X}$, all $a \in V\}$.

Here, Δ is the unit disk in $\mathbb{C}$.

c) Given V with property a), then $\hat{T}(\tilde{\Gamma}) + \lambda.V$ is a bounding set contained in X, all $\lambda < 1$. Furthermore, $\hat{T}(\tilde{\Gamma}) + V$ is contained in $\widehat{T + V}(\tilde{\Gamma})$, whenever Γ is an algebra and $Z = \mathbb{C}$.

Proof.-

a) The set $\Omega = \{f \in \tilde{\Gamma} \mid \|f\|_T \leqslant 1\}$ is a neighbourhood of $0 \in \tilde{\Gamma}$; actually, Ω is a barrel since $\tilde{\Gamma}$ is normal. Now, by strong invariance by derivation, we get some neighbourhood V of $0 \in E$, and some neighbourhood Ω' in $\tilde{\Gamma}$ such that $\|f^n(a)\| \leqslant 1$, all $f \in \Omega'$, all $a \in V$. Thus, the Taylor expension of any $f \in \Omega'$ is uniformly convergent in $\lambda.V$ $(\lambda < 1)$ and $\|f \circ p_x^{-1}[p(x) + a]\| < (1-\lambda)^{-1}$ all $f \in \Omega'$, all $a \in \lambda.V$, all $x \in T$. Since $\tilde{\Omega}$ spans $\tilde{\Gamma}$ and $\tilde{X}$ is the Γ-maximal extension, we have got the property a).

b) There is nothing to be proved, when the right hand of (b) vanishes. Thus, we can take some $r > 0$ such that $T + r.\Delta.a$ is a bounding set contained in X, all $a \in V$. Given $a \in r'.V$, $r' < r$, the first part a) entails the existence of a balanced neighbourhood of $0 \in E$ such that $T + r/r'.\Delta.(a + W)$ is also a bounding set contained in $\tilde{X}$. By Cauchy integral, (proposition 3.2,d), we get :

$$\|f^n(a + \alpha)\|_{\hat{T}} \leqslant \left(\frac{r'}{r}\right)^n \|f\|_{T+r/r'.\Delta.(a+W)}, \text{ all } \alpha \in W, f \in \Gamma .$$

The series $\sum_{n \geqslant 0} f^n(x, a + \alpha)$, $x \in \hat{T}$, defines a function $f_a(\alpha)$ which belongs to $\mathcal{O}_X(W,Z)$; we denote by $(f_a)_\alpha$ the germ defined by f_a at $\alpha \in W$.

When $a' \in a + W$, the following relation exists between $f_{a'}$ and f_a : $(f_{a'})_{\alpha=0} = (f_b)_{\alpha=a'-a}$; hence, the mapping $a \to (f_a)_{\alpha=0}$ is continuous from $r'.V$ into $\tilde{X}$, since $(f_{b=0})_{\alpha=0}$ is the germ of f at x. We must recall the construction of $\tilde{X}$ in chapter I, and (b) is just proved.

c) When $b = 0$, the same computation gives :

$$\|f(x+\alpha)\| \leqslant (1-\lambda)^{-1} \|f\|_{T+V}, \text{ all } \alpha \in \lambda.V, \quad |\lambda| < 1 .$$

When Γ is an algebra, the same inequality for powers of f gives :

$$|f(x+\alpha)| \leqslant (1-\lambda)^{-\frac{1}{k}} \|f\|_{T+V} \text{ all } \alpha \in \lambda.W, \text{ all } k,$$

Then, $\|f\|_{\hat{T}+V} \leqslant \|f\|_{T+V}$, and $\hat{T}(\tilde{\Gamma})+V$ is contained in $\widehat{T+V}(\tilde{\Gamma})$.

Corollary 4.8.- *If X is the maximal extension of some set A in $\mathcal{O}_X(X,Z)$, and A is invariant by derivation, then the following relation exists for any compact set K in X and any balanced, open, neighbourhood W of $0 \in E$:*

$$d_X^W\left[\hat{K}(A)\right] = d_X^W(K) \quad .$$

Remark : Here, W can be chosen such that $K + W$ is contained in X , then $d_X^W(\hat{K}(A)) > 0$, for such W .

Proof.- Since $\hat{K}(A) = \hat{K}(\bar{A})$ and $\bar{A}$ is also invariant by derivation by proposition 3.3.d, we can assume that A is closed for the compact open topology. Now, given $f \in A$, we know by proposition 4.4 that f belongs to a natural, s. i. d, Frechet space Γ ; thus, f belongs to $\Gamma \cap A$ which is also a s. i. d, Frechet space with Γ together , since the Γ-topology is stronger than compact convergence.

Denote as u_Γ the morphism from X to the $\Gamma \cap A$-maximal extension $\tilde{X}_\Gamma$ of X ; then $u_\Gamma\left[\hat{K}(A)\right] + d^W_{\tilde{X}_\Gamma}(K).W$ is contained in $\tilde{X}_\Gamma$ by theorem 4.8.b.

Since X is the intersection of the manifolds $\tilde{X}_\Gamma$ by theorem 1.2, $\hat{K}(A) + d_X^W(K).W$ is contained in X ; then the inequality $d_X^W \hat{K}(A) \geqslant d_X^W(K)$ is proved, and the converse is obvious.

Definition 4.4.- *A sequence (x_n) in X say to reach the boundary whenever $d_X^W(x_n)$ converges near 0 for all balanced neighbourhoods W of 0 in E .*

Theorem 4.9.- *The following properties of the manifold X are equivalent, related to a natural, s. i. d, Frechet space Γ in $\mathcal{O}_X(X,Z)$ which separates X .*

(i) X is the maximal extension of Γ .

(ii) For any sequence (x_n) in X which reaches the boundary, there exists $f \in \Gamma$ with $\sup \|f(x_n)\| = +\infty$.

(iii) If E has a countable basis of open sets, then X is the maximal extension of some $f \in \Gamma$.

Proof.- (i) $\Longrightarrow$ (ii) .

When $\sup_n \|f(x_n)\| < \infty$ all $f \in \Gamma$, the sequence (x_n) is a bounding set for Γ , then theorem 4.8.a) entails (ii).

(ii) $\Longrightarrow$ (i).

Let $\tilde{X}$ be the maximal extension of Γ which its morphism $u : X \to \tilde{X}$. By proposition 1.8, u is injective and so, X is a domain in $\tilde{X}$.

Given a boundary point x_o of X in $\tilde{X}$, then any sequence which converges near x_o in X reaches the boundary of X ; however, this sequence is a bounding set for Γ .

(i) $\Longrightarrow$ (iii).

It is the theorem 2.3 ; the converse implication is obvious.

<u>Definition</u> 4.5.- *Given a finite dimensional, affine, subspace H in E , $p^{-1}(H)$ is a manifold spread over H which will be denoted as X_H . The manifold (X,p) will be called pseudoconvex whenever X_H is a Stein manifold for any H .*
$L(E,Z)$ is the set of continuous endomorphisms from E into Z .

<u>Proposition</u> 4.10.- *If X is the maximal extension of a set A in $\mathcal{O}_X(X,Z)$, and A is invariant by derivation and contains the mappings $u \circ p$ with $u \in L(E,Z)$, then X is pseudoconvex.*

<u>Proof</u>.- Given a compact set K in X_H and $x \in \hat{K}[\mathcal{O}(X_H,\mathbb{C})]$, then for any α in the adjoint space Z' of Z and any $f \in A$, we have ; $|\alpha \circ f(x)| \leq \|\alpha \circ f\|_K$; taking the upperbound of both sides when α describes Z' , we get : x belongs to $\hat{K}[A]$. Then $\hat{K}[\mathcal{O}(X_H,\mathbb{C})]$ is contained in $\hat{K}[A]$ and by corollary 4.8, $\hat{K}[\mathcal{O}(X_H,\mathbb{C})]$ cannot reach the boundary and so X_H is holomorphically convex.

Finally, $L(E,Z)$ separates E and so by proposition 1.9, A separates X ; then given x and x' in X_H , $x \neq x'$, there exists $f \in A$ with $f(x) \neq f(x')$, and by Hahn-Banach theorem there is some $\alpha \in Z'$ with $\alpha \circ f(x') \neq \alpha \circ f(x')$. The manifold X_H is separated by $\mathcal{O}(X_H,\mathbb{C})$, so it is a Stein-manifold.

<u>Comment</u>.- We do not study here the converse implication in proposition 4.10. A result of Gruman and C.O. Kiselman [36,37] , completed by Hervier [41] says that any pseudoconvex manifold (X,p) spread over a Banach space E with a basis is the maximal extension of some $f \in \mathcal{O}_X(X,\mathbb{C})$, and is $\mathcal{O}_X(X,\mathbb{C})$-convex (cf. definition 5.2).

Moreover, an important example of Josephson (Upsala Univ., not published) shows a domain in $E = \ell_\infty(A)$, with A uncountable, which is pseudoconvex with a proper $\mathcal{O}_E(\mathbb{C})$-extension.

Proposition 4.11.- *If X is the maximal extension of some s.i.d., natural, Frechet space Γ in $\mathcal{O}_X(X,Z)$, which separates X, then there exist a countable, admissible covering $\mathcal{U}$ of X such that X is the maximal extension of the associated class $A_{\mathcal{U}}$ in $\mathcal{O}_X(X,\mathbb{C})$.*

Proof.- By corollary 4.4 there exist a such $\mathcal{U}$ such that X is the maximal extension of the associated class $A_{\mathcal{U}}(Z)$ in $\mathcal{O}_X(X,Z)$. Now let us given a sequence (x_n) which reaches the boundary in X and by (ii) in theorem 4.9 a function $f \in A(Z)$ such that $\sup \|f(x_n)\| = +\infty$. The sequence $f(x_n)$ is not weakly bounded in Z', therefore there exists $\zeta \in Z'$ such that $\sup |\zeta \circ f(x_n)| = +\infty$. Since $\zeta \circ f$ belongs to $A_{\mathcal{U}}$ the announced result is a consequence of (ii) in theorem 4.9.

Corollary 4.11.- *With the assumptions of the previous proposition, if more E has a countable basis of open sets, there exists $f \in \mathcal{O}(X,\mathbb{C})$ such that X is f-maximal.*

Proof.- Clear by (iii) in theorem 4.9.

A pathological example.

If the countableness in (iii) is not satisfied by E, X could be the maximal extension of some s.i.d., natural, Frechet space in $\mathcal{O}(X,\mathbb{C})$ and never is f-maximal for all $f \in \mathcal{O}(X,\mathbb{C})$ as the following example will be showing.

Let T be a discrete compact space such that card $T > \aleph_0$, for instance $\{0,1\}^{\mathbb{N}}$, and E the Banach subspace of $C(T,\mathbb{C})$ provided by continuous functions whose supports are countable with uniform norm. The unit ball in E is denoted by ω.

Proposition 4.12.-

a) There exists a proper, direct subspace H of E such that $f = f \circ p$, all $f \in \mathcal{O}(\omega,\mathbb{C})$, for a convenient projection p onto H.

b) Let g be a holomorphic function in the unit disk which has no proper continuation. Then the mapping $\varphi : x \to g \circ x$ belongs to $\mathcal{O}(\omega,E)$

and ω is φ-maximal.

Proof.-

a) Let us given $f \in \mathcal{O}(\omega, \mathbb{C})$; since the mapping $(x_1,\ldots,x_n) \to f^n(0,x_1,\ldots,x_n)$ is n-linear and continuous, there exists $\mu_n \in [\otimes_\pi^n E]'$ such that $f^n(0,x_1,\ldots,x_n) = \mu_n(x_1 \otimes \ldots \otimes x_n)$; here $\otimes_\pi$ is the tensor product equipped with the projective topology.

Let E_n be the Banach subspace of $C(T^n, \mathbb{C})$ provided by continuous functions whose supports are countable and β the n-linear continuous mapping : $(x_1,\ldots,x_n) \to x_1 \ldots x_n$ from E^n into E_n . The universal property of the tensor product provides a topological isomorphism between the completion $\hat{\otimes}_\pi^n E$ and the subspace E_n provided by functions with countable supports, therefore μ_n is the restriction to $\otimes_\pi^n E$ of a measure on T^n . Let us denote by $I_n = \{t \in T^n \ ; \ |\mu_n| (\{t\}) \neq 0\}$, I_n is countable and for any $x \in E$ with support outside I_n , $\mu_n(x) = 0$ since the support of x is countable.

Let I be the union of I_n , K the subspace of E provided by functions whose support does not cut I , H the subspace of E provided by functions whose support is contained in I , we have topologicaly $E = H \oplus K$ and $f^n(0,x) = \int x(t_1) \ldots x(t_n)\, d\mu_n(t) = f^n(0,x_H)$, here x_H is the projection of x onto H . Therefore f has a continuation in $\omega + K$.

b) The map φ is continuous and Gateaux-analytic, thus it belongs to $\mathcal{O}(\omega,E)$. Lastly if ω were not φ-maximal, there might exist $x_0 \in E$ with $\|x_0\|_\infty = 1$ such that $\Theta : \lambda \to g[\lambda\, x_0(t)]$ is holomorphic in a neighbourhood U of 1 for all $t \in T$. When we choose t such that $|x_0(t)| = 1$, we find a contradiction with the choice of g .

Corollary 4.12.- *With the above notation, ω is maximal for a s.i.d., natural Frechet space in $\mathcal{O}(\omega, \mathbb{C})$, but is never f-maximal for all $f \in \mathcal{O}(\omega, \mathbb{C})$.*

Proof.- Clear by proposition 4.10 and proposition 4.5.

Comment.- Another example can be found in [47] .

CHAPTER V : HOLOMORPHIC CONVEXITY

§ 1.- Some Bischop's Lemmas.

Let U be the unit ball of $\mathbb{C}^n$, normed by $\|z\| = \sup |z_i|$ with $z = (z_i)$. The distinguished boundary of U is U^*.

Definition 5.1.- A polynomial P in $\mathbb{C}^n$ is normalized if the maximum value of its coefficients is 1. It is of degree $(d_1, \ldots, d_n)$ if it is of degree at most d_j in the jth variable.

Lemma 5.1.- Given r, $0 < r < 1$, there exists a constant M such that for all t, $0 < t < 1$, the Lebesgue measure of Λ_d defined as $\{z \in r.U \mid |P(z)| \leqslant t^d\}$ is less than $-M/\mathrm{Log}\,t$, for all normalized polynomials P in $\mathbb{C}^n$ of degree $(d, d, \ldots, d)$.

Proof.- Let z^q be the monomial of P whose coefficient is 1.

$$z^q = \int_{U^*} P(z.w)w^{-q}\,\frac{dw}{w} .$$

Therefore, we get $(\frac{r}{2})^{n.d} \leqslant \left| \int_{U^*} P(\frac{r}{2}.w)w^{-q}\,\frac{dw}{w} \right| \leqslant \|P\|_{r/2.U^*}$ and hence there exists $z_o \in r/2\ U^*$ with $|P(z_o)| \geqslant (\frac{r}{2})^{nd}$.

Now, we use the mean value integral in $r.U^*$ and μ_{z_o}, the Poisson kernel of rU^* at z_o. Since $\mathrm{Log}\,\frac{|P|}{(d+1)^n}$ is negative in $r.U$, we obtain

$$nd(\mathrm{Log}\,r - \mathrm{Log}\,2) - n\,\mathrm{Log}(d+1) \leqslant \int \mathrm{Log}\,\frac{|P(w)|}{(d+1)^n}\,d\mu_{z_o}(w) , \text{ and}$$

$$\mu_{z_o}(\Lambda_d) . [d.\mathrm{Log}.\,t - n.\mathrm{Log}\,(d+1)] \geqslant n.d.\,(\mathrm{Log}.r - \mathrm{Log}.2) - n.\mathrm{Log}(d+1) .$$

When z_o describes $r/2.U^*$, the Poisson measure μ_{z_o} is uniformly equivalent with the Lebesgue measure of $r.U^*$.

Then, there exists a constant M' such that

$$\mathrm{mes}\,.\,\Lambda_d \leqslant M' \,.\, \frac{nd(\mathrm{Log}\,r - \mathrm{Log}\,2) - n\,\mathrm{Log}(d+1)}{d\,\mathrm{Log}\,t - n\,\mathrm{Log}(d+1)} .$$

Finally, for a suitable M which only depends on n and r we have

$$\text{mes. } \Delta_d \leqslant - \frac{M}{\text{Log } t} .$$

Lemma 5.2.- Let Q be a compact subset of a manifold (X,p) spread over $\mathbb{C}^n$ and let $f \in \mathcal{O}_X(X, \mathbb{C})$.

There is an $r < 1$ and an integer $e_0 > 0$, such that for all integers d and e with $d \geqslant e \geqslant e_0$, there exists a normalized polynomial P of degree (d, ..., d, e) in n+1 variables such that

$$\|P(p(z),f(z))\|_Q \leqslant r^{d.e^{1/n}} .$$

Proof.- Cover Q by finitely many polydiscs $x_i + \frac{1}{2}\varepsilon U$, where ε is chosen such that $Q + \varepsilon U$ is contained in X, x_i $(1 \leqslant i \leqslant N)$ are in Q. Let c be given such that $c > \sup.\left[1, \|p_i\|_{Q+\varepsilon U}, \|f\|_{Q+\varepsilon U}\right]$.

Let L be the vector space of all polynomials in n+1 variables of degree (d, ..., d, e). For each i, $1 \leqslant i \leqslant N$, and each $k = (k_1, \dots, k_n)$, let ω_{ik} be the linear functional on L defined by $\omega_{ik}(P) = D^k P(p_1, \dots, p_n, f)(x_i)$. Now the dimension of L is $(d+1)^n(e+1)$, and there are fewer than Nt^n functionals ω_{ik} with $|k| < t$; so if t is chosen to be the greatest integer less than $N^{-1}(d+1)(e+1)^{1/n}$, there is a nonzero $P \in L$ such that $\omega_{ik}(P) = 0$ for $|k| < t$, $1 \leqslant i \leqslant N$. We may take P to be normalized.

Let $\tilde{P} = P(p_1, \dots, p_n, f)$. Then by our choice of c we obtain

$$\|\tilde{P}\|_{Q+\varepsilon.U} \leqslant (d+1)^n(e+1)c^{nd+e} ,$$

since P is a sum of at most $(d+1)^n(e+1)$ terms of the form $\alpha p_1^{j_1} \dots p_n^{j_n} f^{j_{n+1}}$ with $|\alpha| \leqslant 1$, $j_i \leqslant d$, $1 \leqslant i \leqslant n, j_{n+1} \leqslant e$. Since $\tilde{P}$ has total order t at each of the points x_i, we have, by Schwarz' lemma in $x_i + \varepsilon.U$

$$|\tilde{P}(x)| \leqslant (d+1)^n(e+1)c^{nd+e} \left|\frac{p(x)-p(x_i)}{\varepsilon}\right|^t .$$

Since Q is contained in $\{x_i + 1/2\, \varepsilon U\}$ we obtain :

$$(1) \qquad |\tilde{P}(x)| \leqslant (d+1)^n(e+1)c^{nd+e}\, 2^{-t} .$$

Now $t+1 \geqslant N^{-1}(d+1)(e+1)^{1/n} \geqslant N^{-1} de^{1/n}$, so the term on the right hand of (1) is dominated by $(r(e))^{de^{1/n}}$, with : $r(e) = (2c)^{n+2/e^{1/n}} . 2^{-N^{-1}}$.

There exists e_o such that $r(e) < 1$ for $d > e > e_o$. We choose $r = r(e_o)$ and we obtain the inequality of the lemma.

<u>Lemma</u> *5.3.- Given a part T of $\mathbb{C}^{n+1}$, an integer e_o such that for all integers d and e with $d \geqslant e \geqslant e_o$, there exists a normalized polynomial $P_{d,e}$ of degree $(d, \ldots, d, e)$ in $\mathbb{C}^{n+1}$ such that :*

$$\|P_{d,e}(z,w)\|_T \leqslant r^{d.e^{1/n}} \text{ , with } r < 1 .$$

Then for all $\varepsilon > 0$, the section T_z of T is finite when z belongs to a subset of $\varepsilon.U$ with nonzero measure.

<u>Proof</u>.- Writing $P_{d,e}(z,w) = \sum_{p \leqslant e} a_p(z)w^p$, some coefficient $a_p(z)$ is a normalized polynomial of degree $(d, \ldots, d)$ which we denote $a_{d,e}(z)$.

Let $\Lambda_{d,e}$ be the set $\{z \in \varepsilon U^* \mid |a_{d,e}(z)| \leqslant r^{1/2d.e^{1/n}}\}$.

By lemma 5.1, for a fixed e , we have for a suitable M :

$$\text{mes}\left[\underline{\lim_{d\to\infty}} . \Lambda_{d,e}\right] \leqslant \left[\overline{\lim_{d\to\infty}} \text{ mes. } \Lambda_{d,e}\right] \leqslant \frac{-M}{e^{1/n}\text{Log } r} .$$

Then, for e large enough, $\underline{\lim}_{d\to\infty} . \Lambda_{d,e}$ has a complement with a nonzero measure. In the following we fix such an e .

Now, for any $z_o \in \{\varepsilon U^* - \underline{\lim}\, \Lambda_{d,e}\}$, $\dfrac{P_{d,e}(z_o,w)}{\sup|a_p(z_o)|}$ is a normalized polynomial of degree $\leqslant e$, denoted by $g_{z_o,d}(w)$. Whenever $(z_o,w) \in T$, $|g_{z_o,d}(w)|$ is less than $r^{1/2.d.e^{1/n}}$.

Since, $g_{z_o,d}(w)$ is normalized with a bounded degree, there exists a sequence d_k such that g_{z_o,d_k} converges to a nonzero polynomial g_{z_o} , and whenever (z_o,w) belongs to T and $z_o \in \{\varepsilon.U^* - \underline{\lim}.\Lambda_{d,e}\}$ we have $g_{z_o}(w) = 0$. That is, T_{z_o} is finite.

§ 2.- Holomorphic convexity.

(X,p) is a manifold spread over the Banach space E .

Definition 5.2.- *Given a subset A in $\mathcal{O}_X(X,\mathbb{C})$; X is called A-convex if $\hat{K}(A)$ is a compact set whenever K is compact set in X .*

If $A = \mathcal{O}_X(X,\mathbb{C})$ and X is pseudoconvex, the aforementionned result of Gruman, C.O. Kiselman, Y. Hervier, says that X is A-convex whenever E has a basis.

Here, we develop the convexity problem when A is a s. i, d, natural Frechet algebra in $\mathcal{O}_X(X,\mathbb{C})$ and X is the maximal extension of A for $\mathcal{O}_E(\mathbb{C})$. The proof does not require the finite dimensional case. Of course, since X is A-maximal, X is pseudoconvex by proposition 4.10 and then X is $\mathcal{O}_X(X,\mathbb{C})$-convex by Gruman's result, but A-convexity is a stronger property.

Let $L(E)$ be the vector space of continuous endomorphisms of E .

We need the following assumptions :

(H_1) For E , there exists a sequence $\pi_k \in L(E)$, with finite dimensional range, which is pointwise convergent to the identity mapping.

(H_2) For A : (i) - the mappings $\xi \circ u_o p$ belong to A for all $\xi \in E'$, all $u \in L(E)$

(ii) - the mappings $x \to f^n(x,u_o p(x))$ belong to A for all $f \in A$, all $u \in L(E)$.

Comment.- The spaces of type $A_{\mathcal{U}}$ fill these requirements whenever $\mathcal{U}$ is a countable, admissible, covering of X such that $p(\omega)$ is bounded, all $\omega \in \mathcal{U}$; also $A_{\mathcal{U}}$ is clearly an algebra.

In the following, K is a compact set in X ; U is the unit ball in E ; X_k is the manifold $p^{-1}[\pi_k(E)]$ which is spread over the finite dimensional vector space $\pi_k(E)$. By theorem 4.8, there exists $\varepsilon > 0$ such that $K+\varepsilon.U$ is contained in X and is a bounding set for A . Furthermore

$$\hat{K}(A) + \alpha.U \text{ is contained in } \widehat{K+\alpha.U}(A) \text{ , all } \alpha \lesssim \varepsilon .$$

The main Lemma

Given a sequence x_n in $\hat{K}(A)$ such that $p(x_n)$ is convergent to $a \in E$: For n large enough, $n \geqslant N$, $\|p(x_n) - p(a)\|$ is less than ε and then there is one point ξ_n in $p^{-1}(a) \cap [x_n + \varepsilon.U]$.

Lemma 5.4.- There exist integers N_1 and $N_2, \varepsilon' > 0$ such that for all $k > N_1$ there exists a compact set Q in X_k such that

$$\left| P\,[\pi_k \circ p(x)\ ,\ f \circ p_x^{-1} \circ \pi_k \circ p(x)] \right| \leqslant \| P\,[\,p(z),\ f(z)]\|_Q$$

for all polynomials P on $\pi_k(E) \times \mathbb{C}$, all $f \in A$, all $x \in \xi_n + \varepsilon'.U$, all $n \in N_2$.

Proof.- By the Banach Steinhaus theorem, $\sup . \|\pi_k\| = M < \infty$; we pick out ε' and ε'' with $0 < \varepsilon' < \varepsilon'' < \frac{\varepsilon}{M+4}$, where ε is chosen such that $\hat{K} + \alpha.U$ is contained in $\widehat{K+\alpha.U}$ all $\alpha \leqslant \varepsilon$, and $K + \varepsilon U$ is a bounding set for A.

On the compact set $p(K)$, π_k is uniformly convergent to the identity, we choose N_1 such that :

(1) $\qquad \|\pi_k(a) - a\| < \varepsilon'$ and $\|\pi_k \circ p - p\|_K < \varepsilon''$, all $k > N_1$.

We fix such a k.

For n large enough, $n \geqslant N_2$, $\xi_n + \varepsilon'U$ is contained in $\hat{K} + \varepsilon''U$ which is itself contained in $\widehat{K+\varepsilon''.U}$.

Let $T = \bigcup(\xi_n + \varepsilon'.U)$ for $n \geqslant N_2$.

By (1) we get $\|\pi_k \circ p - p\|_T \leqslant (M+2)\varepsilon'$ and

$$\|\pi_k \circ p - p\|_{K+\varepsilon''U} \leqslant (M+2)\varepsilon'' .$$

By the Cauchy integral (Prop. 3.2.d), we obtain for all $f \in A$

$$\|f^n(x,\pi_k \circ p(x) - p(x)\|_{T \cup (K+\varepsilon''.U)} \leqslant \left(\frac{M+2}{M+3}\right)^n \|f\|_{\hat{K}+\varepsilon.U}$$

$$\leqslant \left(\frac{M+2}{M+3}\right)^n \|f\|_{K+\varepsilon.U} .$$

We now use the function $g_p(x) = \sum_{n \leqslant p} f^n(x,\pi_k \circ p(x) - p(x))$;

The previous computation gives the uniform convergence of g_p near $f \circ p_x^{-1} \circ \pi_k \circ p(x)$ on $T \cup (K+\varepsilon''U)$.

Now, given a polynomial P on $\pi_k(E) \times \mathbb{C}$, by the assumption (H_2), the mapping $P(\pi_k \circ p, g_p)$ belongs to A, so we get ;

$$\|P[\pi_k \circ p(x), f \circ p_x^{-1} \circ \pi_k \circ p(x)]\|_T = \lim_{p\to\infty} \|P[\pi_k \circ p, g_p]\|_T$$

$$\leq \lim_{p\to\infty} \|P[\pi_k \circ p, g_p]\|_{\hat{K}+\varepsilon''.U}$$ since T is contained in $\hat{K} + \varepsilon''.U$.

Furthermore, since $P[\pi_k \circ p, g_p]$ belongs to A , we have

$$\|P[\pi_k \circ p, g_p]\|_{\hat{K}+\varepsilon''.U} \leq \|P[\pi_k \circ p, g_p]\|_{K+\varepsilon''.U}$$

When p diverges to infinity we get

$$\|P[\pi_k \circ p(x) , f \circ p_x^{-1} \circ \pi_k \circ p(x)]\|_T \leq$$

$$\|P[\pi_k \circ p(x), f \circ p_x^{-1} \circ \pi_k(x)]\|_{K+\varepsilon''.U}$$

Now, we have just to prove that $\{ p_x^{-1} \circ \pi_k \circ p(x)\}$ is a relatively compact set in X_k when x describes $K+\varepsilon''.U$.

Given a sequence x_n in $K+\varepsilon''U$, since K is a compact set, there exists $y \in K$ and a subsequence (x_n) such that $x_n \in y + 2.\varepsilon''.U$. Then, $p_{x_n}^{-1} = p_y^{-1}$ in $p(y) + \varepsilon.U$. Moreover we have previously obtained :
$\|\pi_k \circ p(x_n) - p(x_n)\|$ less than $(M+2)\varepsilon''$ and therefore we have ;
$\|\pi_k \circ p(x_n) - p(y)\| \leq (M+4)\varepsilon'' < \varepsilon$ and :

$$p_{x_n}^{-1} \circ \pi_k \circ p(x_n) = p_y^{-1} \circ \pi_k \circ p(x_n) \quad .$$

Now, since $\pi_k(E)$ is finite dimensional, there exists another subsequence (x_n) such that $\pi_k \circ p(x_n)$ is convergent in $[p(y) + \varepsilon.U] \cap \pi_k(E)$ and the proof is complete.

The convexity theorem.

Theorem 5.1.- Given a s. i. d, natural Frechet algebra A in $\mathcal{O}_X(X,\mathbb{C})$ which satisfies (H_2), where X is a manifold spread over the Banach space E with property (H_1).
If X is the maximal extension of A for $\mathcal{O}_E(\mathbb{C})$, then X is A-convex and separated by A.

Proof. E is separated by the adjoint space E' , so by proposition 1.9

and (H_1), X is separated by A.

Given a compact set K in X, $p(\hat{K}(A))$ is contained in the linear convex hull of $p(K)$ since $\xi \circ p$ belongs to A, $\xi \in E'$. Hence $p(\hat{K}(A))$ is relatively compact.

Given a sequence (x_n) in $\hat{K}(A)$, there exists a subsequence (x_n) such that $p(x_n)$ converges to $a \in E$.

Now, we use the points $\xi_n \in p^{-1}(a) \cap [x_n + \varepsilon'.U]$ introduced in lemma 5.4 for n large enough.

If is there an infinite number of ξ_n which are equal at $y \in X$, then there exists a subsequence (x_n) which converges near y and the proof is over.

We prove by contradiction that this case always happens. Suppose all the ξ_n are distincts. For k large enough, $\pi_k(E)$ cuts $p(a) + \varepsilon'U$. For such a fixed k, the points $\xi_{n,k} = p_{\xi_n}^{-1}[\pi_k(a)]$ are all distinct and there is some $f \in A$ which separates them by proposition 2.4 and by the Baire property of A. Now we consider such an f.

The mappings $u_{n,m} = f \circ p_{\xi_n}^{-1} - f \circ p_{\xi_m}^{-1}$ belong to $\mathcal{O}_E[p(a)+\varepsilon'U, \mathbb{C}]$ and $u_{n,m}[\pi_k(a)] = f(\xi_{n,k}) - f(\xi_{m,k}) \neq 0$. Thus, given a polydisc Δ in $\pi_k(E) \cap [p(a) + \varepsilon'.U]$ not every $u_{n,m}$ vanishes at any point of Δ outside a set Z of measure zero. Thus f takes an infinite number of values on $p^{-1}(z) \cap [\bigcup(\xi_n + \varepsilon'U)]$, whenever $z \in \Delta^* - Z$.

We now prove the converse. By lemma 5.4, for a suitable k and for n sufficiently large, there exists a compact set Q in X_k such that ;

(1) $|P[\pi_k \circ p(x), f \circ p_x^{-1} \pi_k \circ p(x)] \leqslant \|P[p(z),f(z)]\|_Q$, for all polynomials P in $\pi_k(E) \times \mathbb{C}$, all $x \in \xi_n + \varepsilon'U$.

By lemma 5.2, for any $d \geqslant e \geqslant e_o$, we can choose a normalized polynomial of degree $(d, \ldots, d, e)$ in $\pi_k(E) \times \mathbb{C}$ such that

(2) $\|P[p-z), f(z)]\|_Q \leqslant r^{d.e^{1/\dim.\pi_k(E)}}$, with $0 < r < 1$.

Now, we apply lemma 5.3 with $T = \{\pi_k \circ p(x), f \circ p_x^{-1} \pi_k \circ p(x)\}$ when x describes $\bigcup[\xi_n + \varepsilon'.U]$. Thus, f takes a finite number of values on $p^{-1}(z) \cap [\bigcup(\xi_n + \varepsilon'U)]$, when z belongs to a subset of Δ^* with a nonzero measure.

Remark : If the manifold has a finite number of sheaves, its always A-convex when it is the maximal extension of any set A in $\mathcal{O}_X(X, \mathbb{C})$ which contains the mappings $\xi \circ p$, $\xi \in E'$, and is invariant by derivation. Only the first part of the previous proof is necessary using corollary 4.8.

CHAPTER VI : SPECTRUM AND MAXIMAL EXTENSIONS

(X,p) is a manifold spread over a complex c. v. s. E. Let A be a unitary subalgebra of $\mathcal{O}(X, \mathbb{C})$ which is invariant by derivation. We denote by $E' \circ p$ the set of mappings $\{\xi \circ p \mid \xi \in E'\}$.

Since $(\xi \circ p)'(a)$ is the constant $\xi(a)$ which belongs to A, then $(E' \circ p) + A$ is also invariant by derivation.

§ 1.- Manifold structure and spectrum.

Given a linear mapping h from $(E' \circ p) + A$ to $\mathbb{C}$, we denote by $h^n(a)$ the mapping $f \to h\left[f^n(a)\right]$ from $(E' \circ p) + A$ to $\mathbb{C}$. It is convenient to notice that $h^n(a)$ vanishes on $E' \circ p$, all $n > 1$, all $a \in E$. Let $S(A)$ be the set of such h, $h \neq 0$, with the following properties ;

(s_1) - The series $\sum\limits_{n \geqslant 0} h_a^n(f)$ is absolutely convergent for all a in some open, balanced neighbourhood V_h of $0 \in E$, all $f \in A$. Then, we denote by $h_a(f)$ the sum of this series.

(s_2) - For all $a \in V_h$ and $f \in A$, there exists a neighbourhood U of $0 \in E$, such that :

$$\text{Sup.}\ |h_b(f)| < +\infty \quad , \quad b \in (a + U) \cap V_h .$$

(s_3) - There exists $\pi(h) \in E$ (necessarily unique by Hahn-Banach theorem) such that $h(\xi \circ p) = \xi \circ \pi(h)$, all $\xi \in E'$.

(s_4) - The restriction to A of h is a homomorphism.

Comment about (s_3). Any $h \in S(A)$ has a restriction to $(E' \circ p) \cap A$ which is weakly continuous for the pairing $\langle E',E \rangle$. The converse is also true ; any homomorphism from A to $\mathbb{C}$ with (s_1) - (s_2) properties, which is weakly continuous in $(E' \circ p) \cap A$ has a continuation in $S(A)$ by the Hahn-Banach theorem.

Proposition 6.1.- Given $h \in S(A)$, then h_a belongs to $S(A)$ for all $a \in V_h$, and we have

(1) $(h_a)_b = h_{a+b}$, all a and b such that $za + z'b' \in V_h$, for

all z and z' in $\mathbb{C}$, $|z| \leqslant 1$, $|z'| \leqslant 1$.

Proof.- a) The series $\sum_{n \geqslant 0, m \geqslant 0} h_a^n(f)\, h_a^m(g)$ is summable, for all $a \in V_h$, all f and g in A. Thus, we have :

$$h_a(f) \cdot h_a(g) = \sum_{n \geqslant 0} \sum_{p+q=n} h_a^p(f) \cdot h_a^q(g) = \sum_{n \geqslant 0} h\Big(\sum_{p+q=n} f^p(a)\, g^q(a)\Big) =$$

$$= h_a(f.g)$$

So, h_a is a homomorphism, all $a \in V_h$.

b) Given a and b in E with the requirements of the proposition then $za + zb' \in V_h$, all $|z| < r$, $|z'| < r$, for a suitable $r > 1$. Therefore $\sum_{n \geqslant 0} h^n_{za+z'b}(f)$ is the expansion by a series of homogeneous polynomials for the holomorphic function $(z,z') \to h_{z.a+z'.b}(f)$ in the bi-disc $|z| < r$, $|z'| < r$.

The following, relation (2), will be proved below ;

$$(2) \quad f^n(a+b) = \sum_{p+q=n} [f^p(a)]^q(b) .$$

Using this, we get $\sum_{p,q} h\,[[f^p(za)]^q(z'b)]$ as the Taylor expansion of $h_{za+z'b}(b)$ on the aforementioned bi-disc. Hence we get :

$$(h_a)_b(f) = \sum_{q \geqslant 0} \sum_{p \geqslant 0} h\,[[f^p(a)]^q(b)] = h_{a+b}(f) .$$

Now, we prove (2) by the following computation ;

$$f^n(a+z'b) = \sum_{q \leqslant n} z'^q \cdot \int_{|t|=1} f^n(a+\tau.z'.b) \frac{dt}{t^{q+1}}$$

$$= \sum_{q \leqslant n} z'^q \int_{|t|=1} \int_{|\tau|=1} f(\tau.a+\tau.t.z'.b) \frac{dt \,.\, d\tau}{t^{q+1}.\tau^{n+1}} .$$

With the new variable $t' = \tau t$, we get ;

$$f^n(a+z'b) = \sum_{q \leqslant n} z'^q \cdot \int_{|t'|=1} \int_{|\tau|=1} f(\tau.a+t'.z'.b) \frac{dt' \,.\, d\tau}{t'^{q+1}.\tau^{n-q+1}}$$

$$= \sum_{p+q=n} z'^q \, [f^p(a)]^q(b) .$$

The proof of (1) is complete.

c) Given $a \in V_h$, there exists a balanced neighbourhood U' of $0 \in E$ such that $z.a + U'$ is contained in V_h , all $|z| < 1$. Then U' is a suitable V_{h_a} with (s_1) property by (1). Further, if U' is chosen in the aforementioned U of (s_2) associated to h , then h_a has itself the property (s_2) by (1). In fact

$$\sup . |(h_a)_b(f)| = \sup |h_{a+b}(f)| < \infty \ , \quad b \text{ describes } U' .$$

Finally, h_a has property (s_3) since

(3) $\pi(h_a) = \pi(h) + a$, all $a \in V_h$, since $h_a(\xi \circ p) = h(\xi \circ p) + \xi(a)$.

<u>*Proposition*</u> *6.2.- The evaluation $\hat{x}$ at any point $x \in X$ belongs to $S(A)$ and $\pi(\hat{x}) = p(x)$.*

<u>Proof</u>.- We can take for $V_{\hat{x}}$, any balanced open neighbourhood of $p(x)$ where p_x^{-1} is defined. Then $\hat{x}_a = p_x^{-1}[p(x)+a]$ all $a \in V_{\hat{x}}$ and thus (s_2) , (s_3) are obvious.

<u>*Theorem*</u> *6.3.- a) Given $h \in S(A)$; the sets $N_V = \{h_a \mid a \in V\}$ are a basis for a Hausdorff topology on $S(A)$, when h describes $S(A)$ and V the sets V_h with (s_1) - (s_2) properties. This topology is finer than pointwise convergence on A .*

For this topology $(S(A),\pi)$ is a manifold which may not be connected, spread over E . The mapping $x \to \hat{x}$ is a morphism u from X to $S(A)$.

b) Let $\tilde{X}(A)$ be the connected component of $u(X)$. Then, the morphism $u : X \to \tilde{X}(A)$ is the maximal extension of A for $\mathcal{O}_E(\mathbb{C})$ an the function $\tilde{f} : h \to h(f)$ is the extension of f .

c) $\tilde{X}(A)$ is separated by the extended algebra $\tilde{A}$ in $\mathcal{O}[\tilde{X}(A),\mathbb{C}]$.

<u>Proof</u>.- a) By the relation (1) of proposition 6.1, N_V is a neighbourhood of each h_a for any $a \in V$, so we have defined a topology $\mathcal{T}$ on $S(A)$ and yet by (1) the mapping $a \to h_a$ is continuous from V in $(S(A), \mathcal{T})$. Since $\pi(h_a) = \pi(h) + a$, π is continuous and $\{a \to h_a\}$ is its inverse mapping restricted to N_V . Hence π is a local homeomorphism. We prove below that the mapping $a \to h_a(f)$ belongs to $\mathcal{O}_E(V, \mathbb{C})$ and then $\mathcal{T}$ is

finer than pointwise convergence on A since $h_{a=o} = h$ and so $\mathcal{T}$ is a Hausdorff topology. Using proposition 6.2, it is easy to check that u is a morphism from X to $(S(A),\pi)$.

b) Now we prove that $a \to h_a(f)$ belongs to $\mathcal{O}_E(V, \mathbb{C})$. Given $a \in V$, where V is the domain in E associated with (s_1) ; for r small enough and b in E, we get by (1)

$h_{a+tb}(f) = (h_a)_{tb}(f) = \sum_{n \geq 0} t^n h_a[f^n(b)]$ and this series is absolutely convergent for $|t| < r$.

Thus, we have proved the Gateaux-analyticity of the said mapping in V. Further by (s_2), $a \to h_a(f)$ is locally bounded and so belongs to $\mathcal{O}_E(V, \mathbb{C})$ by proposition 3.4. We have just proved that $\tilde{f}$ belongs to $\mathcal{O}(S(A), \mathbb{C})$; indeed $\tilde{f}$ extends f.

Now, we take an A-extension $v : X \to Y$ for $\mathcal{O}_E(\mathbb{C})$ and we denote as v^* the extension mapping from A onto $\mathcal{O}_Y(Y, \mathbb{C})$. With the same arguments as for $\hat{x}$, we can easily see that $y \circ v^*$ belongs to S(A) for any $y \in Y$, and the mapping $w : y \to \hat{y} \circ v^*$ is a morphism from Y to S(A) ; hence the range of w is contained in $\tilde{X}(A)$ since it is connected and contains u(X). Actually : $(x \circ v(x))(f) = (v^*f)(v(x)) = f(x)$, and then $w \circ v = u$ and moreover the maximality of $\tilde{X}(A)$ is proved.

c) is obvious.

§ 2.- Topological spectrum and maximal extensions.

Here, A is a unitary and topological subalgebra of $\mathcal{O}_X(X,\mathbb{C})$ which is invariant by derivation. We obtain some information about the relationship between the topological spectrum sp. A of A and $\tilde{X}(A)$, the last one being defined above without a topology on A.

We denote as $sp.^* A$ the set of $h \in sp. A$ which are weakly continuous on $(E' \circ p) \cap A$. The set of their weak continuation to $A+E' \circ p$ is denoted as $sp.^{**} A$.

Theroem 6.4.- Given A, a natural, s. i. d, Frechet algebra in $\mathcal{O}_X(X, \mathbb{C})$; then $sp.^{**}A$ is contained in S(A) and $\tilde{X}(A)$ is a connected component of $sp.^{**}A$ for the topology induced by S(A) on $sp.^{**}A$.

Proof.- First, we prove that the s. i. d. assumption entails that $sp.^{**}A$ is contained in S(A). Actually, given $h \in sp.^{**}A$, there exists $q \in N(A)$, a balanced neighbourhood V of $0 \in E$, $\varepsilon > 0$, such that :

$|h[f^n(a)]| < 1$, all $f \in A$ with $q(f) < \varepsilon$, all $a \in V$, all integers n.

Then, since $\{f \in A \mid q(f) < \varepsilon\}$ is an absorbing set, the requirements (s_1) and (s_2) are satisfied with $U = V_h = \lambda.V$ $(\lambda < 1)$.

Now, we have only to prove that $\tilde{X}(A)$ is contained in sp. A . Let u be a morphism from X to $\tilde{X}(A)$, we know by theorem 4.7 that the extended Frechet space $u^*(A)$ is also s. i. d. and natural. That is the mapping $f \to \tilde{f}(h)$ is continuous from A to $\mathbb{C}$, for any $h \in \tilde{X}(A)$. Since $f(h) = h(f)$; the proof is complete.

Remark.- When $E' \circ p$ is contained in A , $sp.^{**} A = sp.^* A$.

Proposition 6.5.- [17,80] . *Suppose E normed and let $\mathcal{U}$ be a countable, admissible covering of X , such that p(ω) is bounded all $\omega \in \mathcal{U}$. The maximal extension of the space $A_{\mathcal{U}}$ for $\mathcal{O}_E(\mathbb{C})$ is a connected component of $sp.^* A$ for the topology of theorem 6.3. Whenever E is reflexive, the same result is true for sp. A .*

Proof.- It is a particular case of theorem 6.4 provided by proposition 4.3.

Proposition 6.6.- *Suppose E is a strong dual of a reflexive Frechet space. Given A a natural, s. i. d, Frechet algebra in $\mathcal{O}_X(X,\mathbb{C})$ which contains $E' \circ p$ and is invariant by derivation, then the maximal extension of A for $\mathcal{O}_E(\mathbb{C})$ is a connected component of sp. A for the topology induced by S(A) .*

Proof.- By corollary 4.2, we know that $E'_\beta = E'_A$ and E'_β is reflexive since E is reflexive. Then $sp.^{**} A = sp. A$.

Remark.- When E is finitely dimensional, $\mathcal{O}_X(X,\mathbb{C})$ is a space of type $A_{\mathcal{U}}$ as we have seen in Chapter IV. More is known : $sp.\mathcal{O}(X,\mathbb{C})$ is connected and therefore is the maximal extension [38] .

Now, we are concerned with topologies on the whole space $\mathcal{O}(X) = \mathcal{O}_X(X,\mathbb{C})$ such that $sp.^* \mathcal{O}(X)$ is related with $\tilde{X} = \tilde{X}[\mathcal{O}(X)]$. We denote by $\mathcal{O}(X)_s$ (resp. $\mathcal{O}(X)_c$, resp. $\mathcal{O}(X)_{p.c}$) the pointwise topology on $\mathcal{O}(X)$ (resp. compact topology, resp. precompact topology).

Since pointwise convergence implies uniform convergence on compact sets for every equicontinuous set in $\mathcal{O}(X)$, the set of precompact balanced convex subsets are equal for $\mathcal{O}(X)_s$ and $\mathcal{O}(X)_c$. It will be denoted by

$K(X)$. If E is metrizable, $\mathcal{O}(X)_c$ is complete and any set of $K(X)$ is compact.

Proposition 6.7.- [80,14] . There is a locally convex, Hausdorff, topology on $\mathcal{O}(X)$, denoted by $\mathcal{O}(X)_\varepsilon$ such that :

a) it is the finest one that induces on any $T \in K(X)$ the pointwise convergence.

b) given a c. v. s. G , a linear mapping from $\mathcal{O}(X)_\varepsilon$ to G is continuous iff its restriction at any $T \in K(X)$ is continuous.

Proof.- Let $\mathcal{F}$ be the family of convex, balanced, absorbing sets M in $\mathcal{O}(X)$ such that $M \cap T$ is a neighbourhood of 0 for the restriction of $\mathcal{O}(X)_s$ at T , all $T \in K(X)$.

$\mathcal{F}$ is a basis of a filter and for any $M \in \mathcal{F}$, $\lambda . M$ belongs to $\mathcal{F}$, $(\lambda > 0)$; actually $\lambda M \cap T = \lambda [M \cap T/\lambda] = \lambda [W \cap T/\lambda] = \lambda . W \cap T$, with W a neighbourhood of 0 for $\mathcal{O}(X)_s$, since $T/\lambda \in K(X)$. Therefore $\mathcal{F}$ is a basis of neighbourhoods of 0 for a locally convex topology $\mathcal{O}(X)_\varepsilon$ which is obviously finer than $\mathcal{O}(X)_s$ and so is Hausdorff.

Now, given $T \in K(X)$, $M \in \mathcal{F}$, $a \in T$, then $T - a$ also belongs to $K(X)$ and $(T-a) \cap M$ is a neighbourhood of 0 in the restriction of $\mathcal{O}(X)_s$ at T , that is, $(a + M) \cap T$ is a neighbourhood of a . To prove that $\mathcal{O}(X)_\varepsilon$ is the finest one, we have only b) to check, thus given a balanced convex open set V in G , then $\Phi^{-1}(V)$ belongs to $\mathcal{F}$ and the proof is complete.

The topology $\mathcal{O}_b$. Let $T \in K(X)$ be given and A_T be the space spanned by T and normed by the Minkowski-norm associated with $\hat{T}$. The canonical mapping $A_T \to A_{T'}$ is continuous whenever T is contained in T' , and $K(X)$ is directed by set inclusion. Since every $f \in \mathcal{O}(X, \mathbb{C})$ belongs to some T , the inductive limit of A_T provides a bornological topology on $\mathcal{O}(X)$ which is written $\mathcal{O}(X)_b$.

Proposition 6.8.- We have $\mathcal{O}(X)_b < \mathcal{O}(X)_\varepsilon < \mathcal{O}(X)_{p.c}$. Here, $<$ means "finer than".

Proof.- For the right hand we use proposition 6.6.b) and the equality of pointwise convergence and precompact convergence on any equicontinuous set. For the left hand, we have only to check the continuity of each mapping $A_T \to \mathcal{O}(X)_\varepsilon$. Thus a convergent sequence in A_T is ultimately in a fixed

$T' \in K(X)$ and is pointwise convergent, then it is convergent in $\mathcal{O}(X)_\varepsilon$.

Proposition 6.9.- *Let F a Frechet c. v. s. be given, and $f \in \mathcal{O}_X(X,F)$.*

a) The map $f^* : \xi \to \xi \circ f$ from F'_c into $\mathcal{O}(X)_\varepsilon$ is continuous.

b) If f is locally topological, then f^* is topological on its range.

Proof.- a) By the Banach-Dieudonné theorem, we have only to check the continuity of f^* in any convex balanced equicontinuous set T in F' for the pointwise topology. A such T is mapped by f^* into a set which belongs to $K(X)$ by Ascoli's theorem and $\mathcal{O}(X)_\varepsilon$ induced the pointwise topology on $f^*(T)$, therefore f^* is continuous.

b) Since f is open, f^* is clearly one to one. Let $\xi_\alpha \circ f$ a convergent net for the ε-topology be given, this net is uniformly convergent on any precompact set of X by proposition 6.8. Any compact set in F is homeomorphic with some compact set in X since f is locally topological, then ξ_α is convergent in F'_c .

Proposition 6.10.- *If E is metrizable, $\mathcal{O}(X)_\varepsilon$ is a semi-Montel space, $\mathcal{O}(X)_b$ is barrelled and is the bornological space associated with $\mathcal{O}(X)_c$.*

Proof.- Given a bounded set B in $\mathcal{O}(X)_\varepsilon$, B is also bounded in $\mathcal{O}(X)_c$ by proposition 6.8, thus B is locally uniformly bounded and equicontinuous by proposition 4.5 and Cauchy inequalities. The closed convex balanced hull $\hat{B}$ of B in $\mathcal{O}(X)_c$ belongs to $K(X)$ by Ascoli's theorem, then B is relatively compact in $\mathcal{O}(X)_\varepsilon$, and so $\mathcal{O}(X)_\varepsilon$ is a semi-Montel space. We have just proved that any bounded set in $\mathcal{O}(X)_c$ is bounded in some space A_T , $T \in K(X)$, used to define the topology $\mathcal{O}(X)_b$, then this topology is the bornological topology associated with $\mathcal{O}(X)_c$. Finally, each $T \in K(X)$ is compact in $\mathcal{O}(X)_c$ since $\mathcal{O}(X)_c$ is complete and so the spaces A_T are Banach spaces ; their inductive limit is barrelled.

Proposition 6.11.- *Suppose E normed ; then the following statements are equivalent :*

(i) $\mathcal{O}(X)_\varepsilon$ is bornological

(ii) $\mathcal{O}(X)_\varepsilon$ is barrelled

(iii) E is finitely dimensional.

Proof.- (i) implies (ii) by proposition 6.8.

(ii) implies (iii) :

Given $x_o \in X$. The mapping $f \to f'(x_o,a)$ from $\mathcal{O}(X)_c$ into $\mathbb{C}$ is continuous by Cauchy Integral (proposition 3.2.d).

Now, consider the mapping $\Phi : f \to f'(x_o)$ from $\mathcal{O}(X)_\varepsilon$ into the adjoint normed space E' ; the set $\{f \in \mathcal{O}(X) \mid [[f'(x_o)[[\leqslant 1\}$ is a barrel in $\mathcal{O}(X)_\varepsilon$ by the first remark. Since $\mathcal{O}(X)_\varepsilon$ is barrelled, Φ is continuous.

Now consider the mapping $\psi : \xi \to \xi \circ p$ from E' into $\mathcal{O}(X)_\varepsilon$; the range of a convergent sequence (ξ_n) of E' belongs to $K(X)$ and $\xi_n \circ p$ is convergent in $\mathcal{O}(X)_c$; then ψ is continuous.

Finally, it is clear that $\Phi \circ \psi$ in the identity ; therefore Φ induces a topological isomorphism from $\mathcal{O}(X)_\varepsilon / \Phi^{-1}(0)$ onto E'. By proposition 6.8 the quotient space E' is a semi-Montel space and is finitely dimensional by Riesz's theorem.

(iii) implies (i), since $\mathcal{O}(X)_c$ is a Frechet space whenever E is finitely dimensional.

Theorem 6.12.- Let $u : X \to Y$ be a $\mathcal{O}(X)$-extension for $\mathcal{O}_E(\mathbb{C})$. Then the extension map u^* from $\mathcal{O}(X)_b$ (resp. $\mathcal{O}(X)_\varepsilon$) onto $\mathcal{O}(Y)_b$ (resp. $\mathcal{O}(Y)_\varepsilon$) is topological whenever E is metrizable.

Proof.- a) b-topology [17,48] .

By theorem 6.10 the b-topologies are bornological and we have to verify that the range of any bounded set by u^* and $(u^*)^{-1}$ is bounded for compact open topology. This property is obvious for $(u^*)^{-1}$.

Let T be a bounded set in $\mathcal{O}(X)_c$, so T is contained in a suitable s. i. d, natural Frechet space by proposition 4.5 ; therefore $u^*(T)$ is locally uniformly bounded by corollary 4.6 and so is equicontinuous by Cauchy inequalities. Finally $u^*(T)$ is closed in $\mathcal{O}(Y)_s$ since T is closed in $\mathcal{O}(X)_s$, and by Ascoli's theorem $u^*(T)$ is a compact set in $\mathcal{O}(Y)_c$.

b) ε-topology [80] .

The continuity of $(u^*)^{-1}$ from $\mathcal{O}(Y)_s$ onto $\mathcal{O}(X)_s$ implies that any $T \in K(Y)$ is mapped in a set of $K(X)$; using theorem 6.6.b), we get the continuity of $(u^*)^{-1}$ from $\mathcal{O}(Y)_\varepsilon$ to $\mathcal{O}(X)_\varepsilon$.

Now, let $T \in K(X)$ be given, the first part a) has shown that $u^*(T)$

belongs to $K(Y)$. Since $K(X)$ and $K(Y)$ are compact in $\mathcal{O}(X)_c$ and $\mathcal{O}(Y)_\varepsilon$ by theorem 6.10, u^* restricted to $K(X)$ is continuous since $(u^*)^{-1}$ is continuous. To complete the proof we apply theorem 6.6.b).

<u>Theorem</u> *6.13.- The maximal extension $\tilde{X}[\mathcal{O}(X)]$ is contained in $sp^*[\mathcal{O}(X)_\varepsilon]$, whenever E is metrizable.*

<u>Proof</u>.- Given $h \in \tilde{X}\ [\mathcal{O}(X)]$, then h belongs to $sp^*[\mathcal{O}(\tilde{X})_s]$; therefore h belongs to $sp^*[\mathcal{O}(X)_\varepsilon] = sp^*[\mathcal{O}(\tilde{X})_\varepsilon]$ by the previous theorem.

<u>Corollary</u> *6.13.- Suppose E a Frechet c. v. s. Then $\tilde{X}[\mathcal{O}(X)]$ is contained in $sp\,[\mathcal{O}(X)_\varepsilon]$.*

<u>Proof</u>.- By proposition 6.9, $\mathcal{O}(X)_\varepsilon$ induces on $E' \circ p$ the precompact topology which is the compact topology since E is complete. Then we have just to apply the Arens-Mackey theorem to get $sp^*\mathcal{O}(X)_\varepsilon = sp\,\mathcal{O}(X)_\varepsilon$.

<u>Comment</u>.- There are other topologies on $\mathcal{O}(X)$ which are useful. The main one of them is the Nachbin topology [62] which is used to study the carriers of functionals on $\mathcal{O}(X)$ [18,29] .

Josephson has constructed (not yet published) an open set ω in the Banach space $\ell_\infty(I) = E$, I non countable, which has a proper extension $\tilde{\omega}$ for $\mathcal{O}_E(\mathbb{C})$, and he has found a point $x \in \tilde{\omega} - \omega$ such that the evaluation $\hat{x}$ is not continuous for $\mathcal{O}(\omega)_c$. Then $\mathcal{O}(\omega)_\varepsilon$ can be stricly finer than $\mathcal{O}(\omega)_c$ by the previous theorem 6.12. Moreover the theorem 6.12 is false for compact topology and such ω .

§ 3.- <u>A particular case</u>, $E = \mathbb{C}^I$.

Nothing is known about $sp^*[\mathcal{O}(X)] - \tilde{X}[\mathcal{O}(X)]$ for a general c. v. s. Nevertheless it is possible to complete theorem 6.13 when E is the product space $\mathbb{C}^I$, with I as a general set. We denote by $\mathcal{F}(I)$ the set of finite subsets of I ; for any Λ of $\mathcal{F}(I)$, π_Λ is the projection map from $\mathbb{C}^I$ onto $\mathbb{C}^\Lambda$; we identify $\mathbb{C}^\Lambda$ with the subspace of $\mathbb{C}^I$ defined by (z_i) with $z_i = 0$, $i \in I-\Lambda$. The unit disk of $\mathbb{C}$ is written Δ .

<u>Theorem</u> *6.14.- Let (X,p) be a manifold spread over $\mathbb{C}^I$ and suppose the maximal extension of $\mathcal{O}(X)$ is X . Then there exists $\Lambda_o \in \mathcal{F}(I)$ and a Stein manifold $\dot{X}$ spread over $\mathbb{C}^{\Lambda_o}$ such that X is isomorphic with $\dot{X} \times \mathbb{C}^{I-\Lambda_o}$.*

Proof. We claim the existence of $\Lambda_0 \in \mathcal{F}(I)$ such that for all $x \in X$ there exist neighbourhoods W of x and V of $\pi_{\Lambda_0} \circ p(x)$ in $\mathbb{C}^{\Lambda_0}$ and W is homeomorphic by p with $V + \mathbb{C}^{I-\Lambda_0}$. Clearly this property is satisfied at some given $x_0 \in X$ and now we verify that Λ_0 obtained at x_0 is suitable at any $x \in X$.

Let H be a finite dimensional subspace of $\mathbb{C}^I$ which contains some given $a \in \mathbb{C}^{I-\Lambda_0}$ and such that x and x_0 belong to a connected component X' of $p^{-1}(H)$. By proposition 4.10, X' is a Stein manifold spread over H, hence $-\operatorname{Log} d_a$ is a plurisubharmonic function on X' [54], with $d_a(x') = \sup \{r \geqslant 0 \mid p(x') + r.\Delta$ is homeomorphic by p with a subset of X' which contains $x'\}$. This function vanishes in $W_0 \cap X'$, then d_a vanishes everywhere on X' [54].

Thus the connected component of $p^{-1}\left[p(x) + \mathbb{C}^{I-\Lambda_0}\right]$ which contains x is homeomorphic by p with $p(x) + \mathbb{C}^{I-\Lambda_0}$ by proposition 1.1. This component is written $x + \mathbb{C}^{I-\Lambda_0}$. Let V be a connected neighbourhood of x which is homeomorphic by p with $p(V)$. We claim that p is one-to-one on $W = \bigcup_{y \in V} (y + \mathbb{C}^{I-\Lambda_0})$. Actually, suppose : $p(y') = p(y'')$, $y' \in y_1 + \mathbb{C}^{I-\Lambda_0}$, $y'' \in y_2 + \mathbb{C}^{I-\Lambda_0}$, then $y_2 \in y_1 + \mathbb{C}^{I-\Lambda_0}$, y' belongs to the connected component of $p^{-1}\left[p(y_2) + \mathbb{C}^{I-\Lambda_0}\right]$ which contains y_2 and $y' = y''$ by proposition 1.1. Since p is an open map, W is homeomorphic with $\pi_{\Lambda_0} \circ p(V) + \mathbb{C}^{I-\Lambda_0}$.

Let $(\mathcal{R})$ be the equivalence relation defined on X by $x \equiv x'(\mathcal{R})$ iff $x' \in x + \mathbb{C}^{I-\Lambda_0}$. Each coset is closed then the quotient space $\dot{X}$ is separated. The projection map defined on $\dot{X}$ by $\pi(\dot{x}) = \pi_{\Lambda_0} \circ p(x)$ is continuous and, with the above notations, π restricted to $\dot{W}$ is an homeomorphism onto $\pi_{\Lambda_0} \circ p(V)$. Then $(\dot{X},\pi)$ is a manifold spread over $\mathbb{C}^{\Lambda_0}$, and $\dot{X} \times \mathbb{C}^{I-\Lambda_0}$ is clearly a manifold spread over $\mathbb{C}^I$. Let φ be the canonical imbedding map from X into $\dot{X}$, the map u defined on X by $u(x) = \left[\varphi(x), \pi_{I-\Lambda_0} \circ p(x)\right]$ is an isomorphism from X onto $\dot{X} \times \mathbb{C}^{I-\Lambda_0}$. Clearly $\dot{X}$ is a Stein manifold spread over $\mathbb{C}^{\Lambda_0}$.

Proposition 6.15.- Let X be a manifold spread over a finitely dimensional space. For each $f \in \mathcal{O}(X \times \mathbb{C}^I)$ with I an arbitrary set :

a) There exists a countable subset $\mathbb{N}$ of I such that $f(x,y) = f(x, \pi_{\mathbb{N}}(y))$, all $(x,y) \in X \times \mathbb{C}^I$

b) This set $\mathbb{N}$ is written $\bigcup I_n$, with I_n an increasing sequence of finite sets. Then the sequence $g_n(x,y) = f(x, \pi_{I_n}(y))$ is convergent to f in $\mathcal{O}(X \times \mathbb{C}^I)_\varepsilon$.

Proof.- a) Let $x_o \in X$ be given. By corollary 3.8 there exists a finite subset Λ_o of I and a suitable neighbourhood V of x_o such that $f(x,y) = f(x, \pi_{\Lambda_o}(y))$, for all $x \in V$, $y \in \mathbb{C}^I$. Then, we construct $\mathbb{N}$ by a countable covering of X.

b) The sequence g_n is locally equal to f for n sufficiently large and therefore is precompact in $\mathcal{O}(X \times \mathbb{C}^I)_s$ and is equicontinuous. The definition of ε-topology gives the announced results.

Theorem 6.16.- Let (X,p) be a manifold spread over $\mathbb{C}^I$ such that X is the maximal extension of $\mathcal{O}(X)$. Then $X = sp.\,\mathcal{O}(X)_\varepsilon$.

Proof.- By theorem 6.14, we can suppose $X = \dot{X} \times \mathbb{C}^J$ with $\dot{X}$ a Stein manifold spread over a finitely dimensional space. Let $h \in sp.\mathcal{O}(X)_\varepsilon$ be given. For every finite subset Λ of J, h_Λ defined by $h_\Lambda(g) = h[g(x, \pi_\Lambda(y)]$ is an homomorphism on $\mathcal{O}(X \times \mathbb{C}^\Lambda)$. It is known [38] that h_Λ is an evaluation and so there exists $(x_\Lambda, z_\Lambda) \in \dot{X} \times \mathbb{C}^\Lambda$ such that $h_\Lambda(g) = g(x_\Lambda, z_\Lambda)$. The separation of $\dot{X} \times \mathbb{C}^\Lambda$ by analytic functions implies that $x_\Lambda = x_{\Lambda'}$ and $\pi_{\Lambda'}(z_\Lambda) = z_{\Lambda'}$ for $\Lambda' \subset \Lambda$. Thus, there exists $(x_o,y_o) \in \dot{X} \times \mathbb{C}^\Lambda$ such that $h_\Lambda(g) = g(x_o, \pi_\Lambda(y_o))$, for all $g \in \mathcal{O}(X \times \mathbb{C}^\Lambda)$, all finite subset Λ of I.

Now let $f \in (\dot{X} \times \mathbb{C}^J)$ be given and take the sequence I_n associated with f by proposition 6.15.b). Since h is continuous in $\mathcal{O}(X \times \mathbb{C}^J)_\varepsilon$ we have :

$$h(f) = \lim.\ h\left[f(x, \pi_{I_n}(y)\right] .$$

Moreover, the previous argument gives $h\left[f(x, \pi_{I_n}(y)\right] = f\left[x_o, \pi_{I_n}(y_o)\right]$, using yet proposition 6.15.b) the proof is complete.

Comment.- The theorem 6.16 is proved by V. Aurich [3] for the b-topology by a more difficult way than here. In this way some results have been obtained in [61] by M.C. Matos.

CHAPTER VII : EXTENSIONS OF VECTOR VALUED ANALYTIC MAPPINGS

§ 1.- Vector valued extensions.

Let a complex sequentially complete c. v. s. F and a manifold (X,p) spread over another complex c. v. s. E be given.

Proposition 7.1.- Every $\mathcal{O}_X(X,F)$-extension Y for $\mathcal{O}_E(F)$ is an extension for $\mathcal{O}_E(\mathbb{C})$.

Proof.- Given $f \in \mathcal{O}_X(X,\mathbb{C})$ and $a \in F$, $a \neq 0$. The mapping $f.a$ belongs to $\mathcal{O}_X(X,F)$ and has an extension $\tilde{f}$ in $\mathcal{O}_Y(Y,F)$. For any $\xi \in a^o$, $\xi \circ \tilde{f}$ vanishes and therefore the range of $\tilde{f}$ is contained in $\mathbb{C}.a = a^{oo}$; thus, $\tilde{f} = \bar{f}.a$, with $\bar{f}$ as an extension of f in $\mathcal{O}_Y(Y,\mathbb{C})$.

Theorem 7.1.- [17,48,80]. Suppose E metrizable then the set of extensions of X for $\mathcal{O}_E(\mathbb{C})$ and $\mathcal{O}_E(F)$ are equal.

Proof.- The converse of proposition 7.1 must to be proved.

a) F is a Frechet space

Let $u : X \to Y$ be an extension for $\mathcal{O}_E(\mathbb{C})$ of $\mathcal{O}_X(X,\mathbb{C})$, and $f \in \mathcal{O}_X(X,F)$. For all $\xi \in F'$, the function $\xi \circ f$ has an extension $\overline{\xi \circ f}$ in $\mathcal{O}_Y(Y,\mathbb{C})$. The mapping $\xi \to \overline{\xi \circ f}$ is continuous from F'_c into $\mathcal{O}_Y(Y,\mathbb{C})_s$ by proposition 6.9 and theorem 6.12. Now by the Arens-Mackey theorem, there exists $\bar{f}(y) \in F$ such that $\overline{\xi \circ f}(y) = \xi \circ \overline{f(y)}$, all $\xi \in F'$, thus $\bar{f}$ is weakly analytic and is an extension of f; we have just to apply proposition 3.5.b) and this part of the proof is complete.

b) F is sequentially complete

For each continuous semi-norm p on F, F_p is the space $F/p^{-1}(o)$ normed by p, $\hat{F}_p$ is the completed space, π_p is the imbedding map $F \to \hat{F}_p$. By the first step, there exists $\tilde{f}_p \in \mathcal{O}(Y, \hat{F}_p)$ such that $\tilde{f}_p \circ u = \pi_p \circ f$. Let Ω be the set of points $y \in Y$ for which there exists $\tilde{f}(y) \in F$ such that $\tilde{f}_p(y) = \pi_p \circ \tilde{f}(y)$, for all continuous semi-norms p. We have just to prove $\Omega = Y$. Obviously $\overset{o}{\Omega}$ contains $u(X)$, then we must prove that $\overset{o}{\Omega}$ is closed. Let $y_o \in \overline{\overset{o}{\Omega}}$ be given, there exists $y_1 \in \overset{o}{\Omega}$ and a balanced neighbourhood U of the origin such that $y_1 + U$ is contained in $\overset{o}{\Omega}$ and $y_o \in y_1 + 2U$. Let $a \in U$ be given,

$\tilde{f}_p(y_1 + za) = \sum_{n \geq 0} \tilde{a}_n^p . z^n$ with $\tilde{a}_n \in \hat{F}_p$, the seriesis convergent for $|z| < 2$. But for $|z| < 1$ $\tilde{f}_p(y_1 + za) = \pi_p \circ \tilde{f}(y_1 + za)$. Since F is sequentially complete, using Cauchy integral, we get some $a_n \in F$ such that $\tilde{a}_n^p = \pi_p(a_n)$ for all p and n . Moreover the serie $\sum_{n \geq 0} a_n . z^n$ is convergent in F for all $|z| < 2$ since F is sequentially complete ; then there exists $\tilde{f}(y_1 + za) \in F$ such that $\tilde{f}_p(y_1 + za) = \pi_p \circ \tilde{f}(y + za)$ for all $|z| < 2$. So, the proof is finished.

§ 2.- Application to a functorial property of the maximal extensions.

Let (X,p) and (X',p') be manifolds spread over a Banach space E ; we denote by $u : X \to (\bar{X},\bar{p})$ and $u' : X' \to (\bar{X}',\bar{p}')$ the maximal extensions for $\mathcal{O}_E(\mathbb{C})$.

Theorem 7.2.- [79] . *Let* φ *a locally bi-analytic mapping from* X *to* X' *be given. There exists a locally bi-analytic mapping* $\bar{\varphi}$ *from* $\bar{X}$ *to* $\bar{X}'$ *such that the following diagram is commutative.*

$$\begin{array}{ccc} X & \xrightarrow{\varphi} & X' \\ {\scriptstyle u}\downarrow & & \downarrow{\scriptstyle u'} \\ \bar{X} & \xrightarrow{\bar{\varphi}} & \bar{X}' \end{array}$$

Proof.- $(X, p' \circ \varphi)$ is a manifold spread over E since φ is locally bi-analytic ; on the other hand $u' \circ \varphi$ is a morphism from $(X, p' \circ \varphi)$ to $(\bar{X}', \bar{p}')$ and for each $f \in \mathcal{O}(X')$, its extension $\bar{f}$ to $(\bar{X}',\bar{p}')$ is the extension of $f \circ \varphi$ from $(X, p' \circ \varphi)$ to $(\bar{X}',\bar{p}')$. Moreover the morphism $u' \circ \varphi$ is the maximal extension of the family $(f \circ \varphi)$ when f describes $\mathcal{O}(X')$.

Now let $\overline{p' \circ \varphi}$ the extension of $p' \circ \varphi$ from (X,p) to $(\bar{X},\bar{p})$ be given by theorem 7.1. Let D_x be the differential operator at x ; since $p' \circ \varphi$ is locally bi-analytic, $x \to D_x^{-1}(p' \circ \varphi)$ is an analytic mapping from (X,p) into the Banach space $L(E)$; by theorem 7.1, $D^{-1}(p' \circ \varphi)$ has an extension to $(\bar{X},\bar{p})$ and it is easy to check that $D^{-1}(p' \circ \varphi) \circ D(\overline{p' \circ \varphi})$ is the identity of $L(E)$; then $\overline{p' \circ \varphi}$ is locally bi-analytic by the implicit functions theorem. Thus, $(\bar{X}, \overline{p' \circ \varphi})$ is a manifold spread over E , and the morphism $u : (X, p' \circ \varphi) \to (\bar{X}, \overline{p' \circ \varphi})$ is an extension for the functions $f \circ \varphi$ when f describes $\mathcal{O}(X')$.

Now, using the first part of the proof, there exists a morphism $\bar{\varphi}$

from $(\bar{X}, \overline{p' \circ \varphi})$ to $(\bar{X}', \bar{p}')$ such that $\bar{\varphi} \circ u = u' \circ \varphi$. Since φ is locally bi-analytic, $\bar{\varphi}$ has the same property.

Remark.- To extend the previous result to more general c. v. s. E , it would be interesting to obtain a good substitute for implicit functions theorem. For instance, there are some Frechet spaces with an implicit functions theorem [81] ; for such spaces theorem 7.2 is yet true.

Corollary 7.2.- [48] . The maximal extension for $\mathcal{O}_E(\mathbb{C})$ of manifold (X,p) spread over E is independent of p (up to an isomorphism) when E is a Banach space.

Proof.- Clear with φ a bi-analytic mapping.

§ 3.- Zorn's theorems.

Here we assume that E has Baire property.

Proposition 7.3.- Let u_n be a sequence of continuous homogeneous polynomials with a fixed degree defined on E and F-valued. If u_n is pointwise convergent to a homogeneous polynomial u , then u is continuous.

Proof.- For each $\pi \in N(F)$, the function $\sup_{n \geq 0} . \pi \circ u_n$ is semi-continuous from below and finite everywhere. By Baire theorem, there exists a not empty open set ω and $M \in \mathbb{R}$ such that $\pi \circ u_n(x) \leq M$, for all $x \in \omega$ and all $n \in \mathbb{N}$, therefore $\pi \circ u$ is also bounded on ω . Let $x_0 \in \omega$ be given, for each $h \in E$ the function $z \to u(h + z x_0)$ belongs to $\mathcal{O}(\mathbb{C},F)$, then we have :

$$u(h) = \frac{1}{2\pi i} \int_{|z|=1} u(h+z.x_0) \frac{dz}{z} \quad , \pi \circ u(h) \leq \int_{|z|=1} \pi \circ u(z^{-1}.h + x_0) d(\arg.z)$$

Then for h sufficiently small $\pi \circ u(h) \leq M$. We have proved the continuity of $\pi \circ u$ for each $\pi \in N(F)$ and the proof is complete.

Theorem 7.4.- [7, 68, 85] . Let f be a Gateaux-analytic map from X into F . If f is continuous at some point then f belongs to $\mathcal{O}(X,F)$.

Proof.- Let $\pi \in N(F)$ be given. By proposition 3.2.f), $\pi \circ f$ is continuous in a neighbourhood of each point of continuity. Then we can introduce the not empty open set $W = \{x \in X \mid \pi \circ f \text{ is continuous at } x\}$. We verify that W has a empty boundary, the proof will be complete.

Using a contradiction argument, let x_o be given in the boundary of W. There exists a complex line through x_o such that $d \cap W$ has a boundary point x_1 which belongs to $d \cap X$. We take a sequence (x'_n), $x'_n \in d \cap W$, which converges to x_1.

For all $h \in E$, all $p \in \mathbb{N}$, $f^p(x'_n,h)$ is convergent to $f^p(x_1,h)$ since f is analytic on the subspace $(d \oplus h) \cap X$, moreover $\pi \circ f^p(x'_n)$ is continuous by proposition 3.2.f), then $\pi \circ f^p(x_1)$ is also continuous for all $p \in \mathbb{N}$, by proposition 7.3.

Now we verify the continuity of $\pi \circ f$ at x_1, the contradiction will be complete.

Since the series $\sum_{p \geq 0} f^p(x_1,h)$ is summable for h sufficiently small (prop. 3.3.c), sup. $\pi \circ f^p(x_1)$ is locally semi-continuous from below and finite, then there exists a not empty open set ω and M such that $\pi \circ f^p(x_1,h) \leqslant M$, for all $p \in \mathbb{N}$ and all $h \in \omega$. Now using the same argument as in proposition 7.3, we check the uniformly boundness of $\pi \circ f^p(x_1)$ in a neighbourhood of the origin. That easily implies the uniformly convergence of $\pi(\sum_{p \geq 0} f^p(x_1))$ to $\pi \circ f$ in a neighbourhood of x_1, and lastly the continuity of $\pi \circ f$ at x_1.

Remark.- This result can be extended from Baire spaces to some other spaces as, for instance, adjoint of Frechet-Schwartz spaces [46].

Theorem 7.5.- Let Λ be a set in $\mathcal{O}_Y(Y,\mathbb{C})$ which separates Y. Let f be a Gateaux-analytic map from X into Y such that :

(i) All $x_o \in X$, there exists a connected neighbourhoods U_{x_o} of the origin in F and ω_{x_o} of x_o such that $f(x) + U_{x_o}$ is contained in Y, for all $x \in \omega_{x_o}$.

(ii) For every $g \in \Lambda$, $g \circ f$ is Gateaux-analytic in X.

(iii) f is continuous at some point of X.

Then f belongs to $\mathcal{O}_X(X,Y)$.

Proof.- $\pi \circ f$ belong to $\mathcal{O}_X(X,F)$ by theorem 7.3, therefore $\pi^{-1}_{f(x)} \circ \pi \circ f$ is analytic in a neighbourhood of each $x \in X$, defines a germ denoted by $\tilde{f}_x$. Hence $W = \{x \in X \mid f_x = \tilde{f}_x\}$ is an open set which is the set of point $x \in X$ such that f is continuous at x. We have just to prove that $W = X$, that is W is closed. Let $x_o \in \bar{W}$ be given, and V an neighbourhood of $0 \in F$ such that $V + V \subset U_{x_o}$.

Since $\pi \circ f$ is continuous, we can choose ω_{x_o} such that : $\pi \circ f(x) \in \pi \circ f(x_o) + V$, for all $x \in \omega_{x_o}$. Furthermore there exists $x_1 \in \omega_{x_o} \cap W$ such that $\pi \circ f(x_o) + V$ is contained in $\pi \circ f(x_1) + U_{x_o}$. Thus $\pi^{-1}_{f(x_1)} \circ \pi \circ f$ belongs to $\mathcal{O}(\omega_{x_o}, Y)$.

On the other hand $g \circ f$ belongs to $\mathcal{O}_X(X, \mathbb{C})$ by theorem 7.3, therefore $g \circ \pi^{-1}_{f(x_1)} \circ \pi \circ f$ and $g \circ f$ are analytic in ω_{x_o} and equal in a neighbourhood of x_1. Hence $g \circ \pi^{-1}_{f(x_1)} \circ \pi \circ f$ and $g \circ f$ are equal in ω_{x_o} and by the separation assumption of Λ, we obtain $f_{x_o} = \tilde{f}_{x_o}$.

Remark.- When Y has a finite number of sheaves over each point of F, the requirement (i) is always satisfied.

§ 4.-Functorial properties of extensions.

(X,p) and (Y, π) are manifolds spread over Frechet spaces E and F.

Let A be an unitary subalgebra of $\mathcal{O}(Y)$. We denote by $u : X \to \tilde{X}$ and $v : Y \to \tilde{Y}(A)$ the canonical morphisms from X into the $\mathcal{O}(X)$-maximal extension and from Y into the A-maximal extension for $\mathcal{O}_E(\mathbb{C})$ and $\mathcal{O}_F(\mathbb{C})$. Let $\varphi \in \mathcal{O}_X(X,Y)$ be given, this section is concerned by the existence of $\tilde{\varphi} \in \mathcal{O}_{\tilde{X}}(\tilde{X}, \tilde{Y}(A))$ such that the next diagram is commutative

$$\begin{array}{ccc} X & \xrightarrow{\varphi} & Y \\ u \downarrow & (1) & \downarrow v \\ \tilde{X} & \xrightarrow[\tilde{\varphi}]{} & \tilde{Y}(A) \end{array}$$

Of course, whenever A is smaller in $\mathcal{O}(Y)$ this problem is easier. It is convenient to notice that problem is solved by theorem 7.2 if φ is a local isomorphism, E and F are Banach spaces.

We denote by φ^* the map $f \to f \circ \varphi$ from $\mathcal{O}(Y)$ into $\mathcal{O}(X)$.

For any $\tilde{x} \in \tilde{X}$, we denote by ${}^t\varphi^*(\tilde{x})$, the map $f \to \widetilde{f \circ \varphi}\,(\tilde{x})$, with ~ is the extension from X to $\tilde{X}$.

Proposition 7.6.- a) ${}^t\varphi^*(\tilde{x})$ *is an homomorphism from the algebra* $\mathcal{O}(Y)$ *into* $\mathbb{C}$

b) $v \circ \varphi = {}^t\varphi^* \circ u$

c) For any $\tilde{x} \in \tilde{X}$, ${}^t\varphi^*$ *satisfy the requirement* (s_3)

in [VI, 1], *that is :*
there exists $b \in F$ *such that* ${}^t\varphi^*(\tilde{x})\ (\xi \circ \pi) = \xi(b)$, *for all* $\xi \in F'$.

<u>Proof</u>.- Properties a) and b) are obvious. By theorem 7.1, $\pi \circ \varphi$ can be extended according to $\widetilde{\pi \circ \varphi}$ which belongs to $\mathcal{O}(\tilde{X},F)$. Hence, we obtain :

$${}^t\varphi^*(\tilde{x})\ (\xi \circ \pi) = \widetilde{\xi \circ \pi \circ \varphi}\ (\tilde{x}) = \xi\ [\widetilde{\pi \circ \varphi}\ (\tilde{x})] \quad .$$

In order to prove that $\tilde{\varphi} = {}^t\varphi^*$ is a solution for the diagram (1) , we should verify that properties (s_1) and (s_2) in (VI,1) must be satisfied by ${}^t\varphi^*$. Unfortunatly, this problem is not yet solved for $A = \mathcal{O}(Y)$ or a general subalgebra A .

<u>*Theorem*</u> *7.6.- The map* ${}^t\varphi^*$ *solves the diagram (1) , if* A *is a s. i. d, natural, Frechet subalgebra of* $\mathcal{O}(Y)$.

<u>Proof</u>.- a) <u>The range of ${}^t\varphi^*$ is contained in $S(A)$</u>

The range of the restriction of φ^* to A is algebrically isomorphic with $A/\text{Ker}\,\varphi^*$. Since A is natural, $\text{Ker}\,\varphi^*$ is closed in A and $A/\text{Ker}\,\varphi^*$ is a Frechet space for the quotient topology. Then $\varphi^*(A)$ with the carried topology of $A/\text{Ker}\,\varphi^*$ is a natural Frechet algebra in $\mathcal{O}(X)$. Since E is metrizable, the extended Frechet algebra $\widetilde{\varphi^*(A)}$ is also natural by theorem 4.7.

Thus, we have proved that the map $f \to \widetilde{f \circ \varphi}\ (\tilde{x})$ is continuous for the topology of A , that is ${}^t\varphi^*(\tilde{x})$ belong to sp.A and further to sp^*A by proposition 7.6. Now, using theorem 6.4, we obtain the announced result.

b) <u>${}^t\varphi^*$ belong to $\mathcal{O}(\tilde{X}, S(A))$</u>

Since $\widetilde{\varphi^*(A)}$ is a natural Frechet algebra, we know by theorem 2.2 that $\widetilde{\varphi^*(A)}$ is locally uniformly bounded and $\tilde{x}_0 \in \tilde{X}$ being given, there exists a neighbourhood ω of $\tilde{x}_0$ such that $\tilde{g} \to \|\tilde{g}\|_\omega$ is a continuous semi-norm for $\widetilde{\varphi^*(A)}$.

On the other hand, the maps $(a,f) \to f^n(a)$ from $F \times A$ into A are equicontinuous since A is s. i. d. and $\widetilde{\varphi^*}$ is a continuous map from A into $\varphi^*(A)$ as it has been proved at the first step. Therefore the maps $(a,f) \to f^n(a) \circ \varphi$ are equicontinuous. Hence, the previous arguments together imply that $\sum_{n \geq 0} \|\widetilde{f^n(a) \circ \varphi}\|_\omega < \infty$ for all a in a convenient neighbourhood U of the origin of F and for all $f \in A$. We have just proved that ${}^t\varphi^*(\tilde{x}) + U$ is contained in $S(A)$ for all $\tilde{x} \in \omega$; that is the assumption (i) of theorem 7.5.

We denote by π the projection of $S(A)$ into F, we have proved in proposition 7.6 that $\tilde{\pi}[{}^t\varphi^*(\tilde{x})] = \widetilde{\pi \circ \varphi}(\tilde{x})$ and $\widetilde{\pi \circ \varphi} \in \mathcal{O}(\tilde{X},F)$. Then the assumption (**iv**) of theorem 7.5 is satisfied by ${}^t\varphi^*$.

$S(A)$ is separated by the extension $\tilde{A}$ and for any $\tilde{g} \in \tilde{A}$ we have $\tilde{g}[{}^t\varphi^*(\tilde{x})] = \widetilde{g \circ \varphi}(\tilde{x})$. Thus the assumption (ii) of theorem 7.5.

Lastly, ${}^t\varphi^*$ is continuous in $u(X)$ by proposition 7.6.**b**). All the assumptions of theorem 7.5 are satisfied by ${}^t\varphi^*$ and the proof is complete after using **an obvious** connectedness argument with theorem 6.4.

<u>*Corollary*</u> *7.6.a.- If A is a space of type $A_{\mathcal{U}}$, which $\mathcal{U}$ as an admissible and countable covering of Y, then $\tilde{\varphi} = {}^t\varphi^*$ solves the diagram (1).*

<u>*Corollary*</u> *7.6.b.- If F is finite dimensional and $A = \mathcal{O}(Y,\mathbb{C})$, then the same conclusion is true.*

They are particular cases of theorem 7.6.

<u>*Theorem*</u> *7.7.- If F is a Banach space with a Schauder basis and $A = \mathcal{O}(Y)$, then the same conclusion is true.*

<u>Proof</u>.- By the Gruman-Kiselman-Hervier theorem $\tilde{Y}(A)$ is the maximal **exten**sion of some $f \in \mathcal{O}(Y)$. By theorem 4.5, f belongs to a space of type $A_{\mathcal{U}}$ and we can apply corollary 7.6.a.

<u>Comment</u>.- The theorem 7.6 can be generalised [48] to maximal extension $\tilde{Y}(A)$ such that for any sequence (y_n) in $\tilde{Y}(A)$ which reach the boundary, there exists $\tilde{g} \in \mathcal{O}[\tilde{Y}(A)]$ such that $\sup |\tilde{g}(y_n)| = +\infty$. If A is a s. i. d. natural Frechet space, this assumption is satisfied by theorem 4.9, but yet no other example does exist.

§ 5.- <u>Extension of products</u>.

Let X and Y be two manifolds respectively spread over c. v. s. E and F. For any $f \in \mathcal{O}(X \times Y)$, f^* is the map $x \to \{y \to f(x,y)\}$ from X into $\mathcal{O}(Y)$.

<u>*Theorem*</u> *7.8.- If E (resp. E and F) is metrizable $f^* \in \mathcal{O}[X, \mathcal{O}(Y)_c]$ (resp : $\mathcal{O}[X, \mathcal{O}(Y)_b]$ and the map $f \to f^*$ is onto $\mathcal{O}[X, \mathcal{O}(Y)_c]$).*

Proof.- To begin with, we show that $(f^n)^*(x;h)$ is a continuous homogeneous polynomial from E into $\mathcal{O}(Y)_c$ (resp. $\mathcal{O}(Y)_b$) for all $n \in \mathbb{N}$, all $x \in X$.

Let (h_i) be a convergent sequence in E to zero, since E is metrizable there exists a sequence ε_i in $\mathbb{R}$ such that $\lim \varepsilon_i = +\infty$ and $\lim \varepsilon_i h_i = 0$. Then $(f^n)(x,\varepsilon_i h_i)$ is bounded in $\mathcal{O}(Y)_c$ (resp. $\mathcal{O}(Y)_b$ by theorem 6.10). Using $\lim \varepsilon_i = \infty$ and homogeneous property, we obtain the continuity of $(f^n)^*(x)$ firstly at zero and after everywhere since $(f^n)^*(x)$ is a polynomial.

Now, ρ being given $(\rho > 1)$ and using Cauchy inequalities, we obtain the following majorization for all $N \in \mathbb{N}$ and all h sufficiently small :

$$\rho^N \left| f^*(x+h)(y) - \sum_{n=0}^{n=N} (f^n)^*(x;h)(y) \right| \leqslant \sup_{|t|=\rho} |f(x+th,y)| .$$

Then, the left hand stays in a bounded set of $\mathcal{O}(Y)_c$ (resp. $\mathcal{O}(Y)_b$), when h describes the sequence (h_i) and for all N.

Since ρ^N tends to infinity, the series $\sum_{n\geqslant 0} (f^n)^*(x,h)$ is uniformly convergent to $f^*(x+h)$ in $\mathcal{O}(Y)_c$ (resp. $\mathcal{O}(Y)_b$) when h describes the sequence (h_i).

With the first step together, we obtain : $f^* \in \mathcal{O}[X,\mathcal{O}(Y)_c]$ resp. $\mathcal{O}[X,\mathcal{O}(Y)_b]$.

Let $g^* \in \mathcal{O}[X,\mathcal{O}(Y)_c]$ be given, we have to prove that $g(x,y) = g^*(x)(y)$ defines an analytic function on $X \times Y$, whenever E and F are metrizable.

Obviously g is separately an analytic function of each variable x and y, then g is analytic on each finite dimensional subspace of $X \times Y$ by Hartogs theorem [8]. Then g is a Gateaux-analytic function.

Now, we verify the continuity and we take two sequences (x_n) and (y_n) which converge to a and b in X and Y. Since g^* is continuous $g(x_n,y_k)$ is uniformly convergent to $g(x_0,y_k)$ when y_k describes the sequence (y_n). Thus we obtain the continuity of g since $g(x_0)$ is continuous.

Corollary 7.8.- If E and F are metrizable, we have $\mathcal{O}(X, \mathcal{O}(Y)_b) = \mathcal{O}(X, \mathcal{O}(Y)_c) = \mathcal{O}[X, \mathcal{O}(Y)_\varepsilon]$.

Comment.- When F is not metrizable the map $f \to f^*$ cannot be onto $\mathcal{O}[X, \mathcal{O}(Y)_c]$. For instance, suppose $X = E$ with E and B an infinite dimensional Banach space and $Y = E'_\sigma$ the weak adjoint space. The duality $\langle E, E'_\sigma \rangle$ defines a continuous linear map from E into $(E'_\sigma)'_c$ but the pairing is not continuous on $E \times E'_\sigma$.

<u>Theorem</u> *7.9.- If E and F are metrizable, the maximal extension of $X \times Y$ for $\mathcal{O}_{E\times F}(\mathbb{C})$ is the product of maximal extensions $\tilde{X}$ and $\tilde{Y}$ of X and Y for $\mathcal{O}_E(\mathbb{C})$ and $\mathcal{O}_F(\mathbb{C})$.*

<u>Proof</u>.- Let π_X (resp. π_Y) be the projection map $X \times Y \to X$ (resp. Y), and V_X (resp. V_Y) the extension morphism $X \to \tilde{X}$ (resp. $Y \to \tilde{Y}$) for $\mathcal{O}_E(\mathbb{C})$ (resp. $\mathcal{O}_F(\mathbb{C})$). To begin with, we verify that $(V_X \circ \pi_X , V_Y \circ \pi_Y)$ is an extension of $X \times Y$ for $\mathcal{O}_{E\times F}(\mathbb{C})$.

Let $\Phi(X,Y)$ the bijective mapping from $\mathcal{O}(X \times Y)$ onto $\mathcal{O}[X,\mathcal{O}(Y)_\varepsilon]$ defined by theorem 7.8 and corollary 7.8. After pointing out $\mathcal{O}(Y)_\varepsilon$ as a sequentially complete space by theorem 6.10, we can introduce the extension map $I : \mathcal{O}[X,\mathcal{O}(Y)_\varepsilon] \to \mathcal{O}[\tilde{X},\mathcal{O}(Y)_\varepsilon]$ given by theorem 7.1 and the topological isomorphism J between $\mathcal{O}(Y)_\varepsilon$ and $\mathcal{O}(\tilde{Y}_\varepsilon)$ given by theorem 6.12.

Then the extension map from $\mathcal{O}(X \times Y)$ onto $\mathcal{O}(\tilde{X} \times \tilde{Y})$ is given by : $\Phi^{-1}(\tilde{X},\tilde{Y}) \circ J \circ I \circ \Phi(X,Y)$.

Now we verify the maximality of $\tilde{X} \times \tilde{Y}$. Let $u : (X \times Y) \to (Z,\pi)$ be an extension for $\mathcal{O}_{E\times F}(\mathbb{C})$.

We introduce the following equivalence relation $\mathcal{R}_E$ (resp. $\mathcal{R}_F$) on Z : $z \equiv z'$ $(\mathcal{R}_E)$ if z' belongs to the connected component of $\pi^{-1}[\pi(z) + F]$ which contains z . Each coset is closed in Z , then $Z/\mathcal{R}(E)$ is a Hausdorff and connected manifold spread over E . Let π_E the imbedding morphism $Z \to Z/\mathcal{R}(E)$.

Let $f \in \mathcal{O}(X)$ be given, then $u^* \circ \pi_X^*(f)$ belongs to $\mathcal{O}(Z)$, is constant on each coset for $\mathcal{R}(E)$ and so defines an analytic function on $Z/\mathcal{R}(E)$ which extends f . The morphism π_E has been constructed such that $\pi_E \circ u$ can be factored through π_X by π'_X such that $\pi'_X \circ \pi_X = \pi_E \circ u$. Then π'_X is an extension of X for $\mathcal{O}_E(\mathbb{C})$, so it can be factoried through $\tilde{X}$ by a morphism $u_X : Z/\mathcal{R}(E) \to \tilde{X}$ which verifies : $u_X \circ \pi'_X = V_X$. Thus we obtain the following commutative diagram :

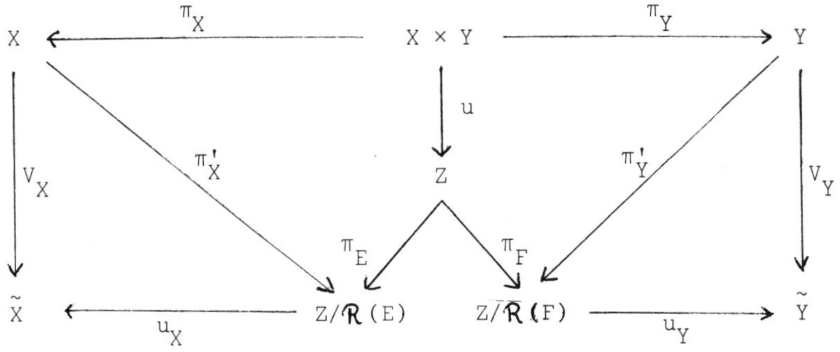

Now, it is easy to verify that $\psi = (u_X \circ \pi_E , u_Y \circ \pi_F)$ is a morphism from Z to $(\tilde{X},\tilde{Y})$ which satisfies $\psi \circ u = (V_X \circ \pi_X , V_Y \circ \pi_Y)$. Then $\tilde{X} \times \tilde{Y}$ is the maximal extension since it can be factored through Z .

CHAPTER VIII : POLYNOMIAL APPROXIMATION

§ 1.- Hilbertian operators

We denote by $H_1 \xrightarrow{A} H_2$ a closely, densely defined linear operator from the Hilbert space H_1 to another H_2 ; the range of A is written R_A and D_A is the everywhere dense subset where A is defined.

The transposed operator of A is denoted by $\tilde{A}$, it is closely, densely defined from H_2 to H_1 ; the restriction of A to $\bar{R}_{\tilde{A}} \cap D_A$ is denoted by $A^{\div}$.

The scalar product of some Hilbert space is written $< , >$ and $\|\cdot\|$ is the associated norm.

The next Von Neumann theorem will be used : $Id + A \circ \tilde{A}$ is one-to-one from $D_{\tilde{A}}$ onto H_2 .

Proposition 8.1.- $A^{\div}$ is one-to-one onto R_A .

Proof.- The said relation $\bar{R}_{\tilde{A}} = (\text{Ker } A)^{\circ}$, which is easily verified, implies that the projection of any $x \in A^{-1}(y)$ on $\bar{R}_{\tilde{A}}$ yet belongs to $A^{-1}(y)$ and it is the one in $A^{-1}(y) \cap \bar{R}_{\tilde{A}}$.

Proposition 8.2.- Let $H_1 \xrightarrow{A} E$ and $H_2 \xrightarrow{B} E$ be given suppose :

(1) $D_{\tilde{A}} \subset D_{\tilde{B}}$ and $\|\tilde{B}(y)\| \leqslant \|\tilde{A}(y)\|$, all $y \in D_{\tilde{A}}$.

Then we have :

(2) $R_B \subset R_A$ and $\|(A^{\div})^{-1}(y)\| \leqslant \|(B^{\div})^{-1}(y)\|$, all $y \in R_B$.

Proof.- Let $y \in R_B$ be given and let $x = (B^{\div})^{-1}(y)$. We are doing with the sequence (y_k) defined in E by $y = \frac{1}{k} y_k + A \circ \tilde{A}(y_k)$, this sequence exists by Von Neumann theorem. We get :

$$\frac{1}{k} \|y_k\|^2 + \|\tilde{A}(y_k)\|^2 = < y,y_k > = < B(x),y_k > = < x,\tilde{B}(y_k) > \leqslant$$
$$\|x\| \, \|\tilde{A}(y_k)\| \text{ by } (1) .$$

An easy computation about $\|A(y_k)\|$ with the previous inequality gives the following estimations :

$$\|\tilde{A}(y_k)\| \leq \|x\| \quad \text{and} \quad \|y_k\| \leq \sqrt{k} \quad \|x\| \quad .$$

Since a closed, bounded set in H_1 is weakly compact, there exists $x_0 \in H_1$, $\|x_0\| \leq \|x\|$, such that x_0 is in the weak closure of $\bigcup_{k \geq n} \tilde{A}(y_k)$ for any n . However $A \circ \tilde{A}(y_k)$ is strongly convergent to y then (x_0,y) belongs to the weak closure of a bounded set of the graph of A . Since this graph is closed we have $y = A(x_0)$ and R_B is included in R_A .

The same argument proves that x_0 belongs to $\bar{R}_{\tilde{A}}$, therefore $y = A^{\div}(x_0)$ and $\|(A^{\div})^{-1}(y)\| \leq \|x_0\| = \|(B^{\div})^{-1}(y)\|$.

Theorem 8.3.- Let $X \xrightarrow{T} Y \xrightarrow{S} Z$ *and* $H \xrightarrow{J} Y$ *be given with* $S.T=0$. *Suppose :*

(3) $D_{\tilde{T}} \cap D_S \subset D_{\tilde{J}}$ *and* $\|\tilde{J}(y)\|^2 < \|\tilde{T}(y)\|^2 + \|S(y)\|^2$, *all* $y \in D_{\tilde{T}} \cap D_S$.

Then we have :

a) $R_J \subset R_T \oplus R_{\tilde{S}}$ *and* $R_T \cap R_{\tilde{S}} = \{0\}$

b) *all* $y \in R_J$, *there exist only one* $u \in D_T \cap \bar{R}_{\tilde{T}}$ *and* $v \in D_{\tilde{S}} \cap \bar{R}_S$ *such that :*

(4) $T(u) + \tilde{S}(v) = y$ *and* $\|u\|^2 + \|v\|^2 \leq \|\tilde{J}(y)\|^2$.

c) *Furthermore if* $y \in Ker.S$ *then* $v = 0$.

Proof.- Let $X \quad Z \xrightarrow{A} Y$ be defined by $A(x,z) = T(x) + \tilde{S}(z)$. Clearly A is densely defined.

Let a sequence (x_n,z_n) such that (x_n,y_n) converges to (x,z) and $T(x_n) + \tilde{S}(z_n)$ converges to y . Since $S \circ T = 0$, we easily verify that $\langle R_T , R_{\tilde{S}} \rangle = 0$, let p be the projection map onto $\bar{R}_T$. The operators T and $\tilde{S}$ are closed, $T(x_n)$ converges to $p(y)$ and $\tilde{S}(z_n)$ converges to $y - p(y)$, then we have $T(x) = p(y)$ and $\tilde{S}(z) = y - p(y)$, that is, A is closed.

Now we verify that the requirement (1) of proposition 8.1 is satisfied by A . We have $D_{\tilde{A}} = D_{\tilde{T}} \cap D_S$ and $\tilde{A}(y) = (\tilde{T}(y), S(y))$, then (3) gives (1) .

Since $\langle R_T , R_{\tilde{S}} \rangle = 0$, we have $\bar{R}_{\tilde{A}} = (\text{Ker } A)^o = (\text{Ker } T \times \text{Ker } \tilde{S})^o = \bar{R}_{\tilde{T}} \times \bar{R}_S$. Then a) and b) are given by (2) .

Lastly, if $y \in \text{Ker } S$ we have $S \circ \tilde{S}(v) = 0$ since $S \circ T = 0$. Then $\tilde{S}(v)$ belongs to $\text{Ker } S \cap R_{\tilde{S}} = \{0\}$ and v belongs to $\text{Ker } \tilde{S} \cap \bar{R}_S = \{0\}$.

§ 2.- A Hörmander's result

Let X be a finitely dimensional Stein manifold, not necessarily spread. We take a relatively compact open set ω in X such that ω is also a Stein manifold, and ω is endowed with a hermitian, Riemannian metric whose invariant element of volume is denoted by $d\sigma$. (cf. [50] for this construction).

For any $\varphi \in C^\infty(\omega)$, $L^2_p(\varphi)$ is the Hilbert space of square integrable differential forms of type $(p,0)$ for the measure $e^{-\varphi} . d\sigma$. The differential operator of Dolbeaut cohomology is written $\bar{d}$.

Let $\alpha = (\alpha_i) \in \mathcal{O}(\bar{\omega})^N$ be given for some integer N . We use the following Hilbertian, densely, closely, defined operators :

$$T^\alpha_N(u) = \Sigma\, \alpha_i\, \bar{d}u_i \quad \text{from } [L^2_0(\varphi)]^N \quad \text{into } L^2_1(\varphi)$$
$$S^\alpha_N(v) = (\alpha_i\, \bar{d}v) \quad \text{from } L^2_1(\varphi) \quad \text{into } [L^2_2(\varphi)]^N$$
$$J^\alpha_N(w) = (\Sigma\, |\alpha_i|^2)^{1/2} . w \quad \text{from } L^2_1(\varphi) \quad \text{into } L^2_1(\varphi) .$$

Lemma 8.4.- The metric of ω and φ can be chosen such that the equation (1) : $T^\alpha_N(u) = v$, has a unique solution in $D_T \cap \overline{R_{\tilde{T}}}$ for any $v \in L^2_1(\varphi)$ such that :

$$\bar{d}v = 0 \quad \text{and} \quad \int |v|^2\, (\Sigma|\alpha_i|^2)^{-1}\, d\sigma < \infty \quad .$$

Moreover, the solution u verifies the following estimation :

$$(2) \qquad \int |u|^2\, e^{-\varphi}\, d\sigma \leq \int |v|^2\, (\Sigma|\alpha_i|^2) e^{-\varphi}\, d\sigma \quad .$$

Proof.- For $N = 1$ and $\alpha = 1$, this lemma is the theorem 5.2.4 of [50] , which gives the following estimation :

(3) $\|v\|^2 \leq \|\tilde{T}(v)\|^2 + \|S(v)\|^2$, all $v \in D_{\tilde{T}} \cap D_S$ with $T = T^1_1$, and $S = S^1_1$.

Using the assumption $\alpha \in \mathcal{O}(\bar{\omega})^N$, we see that $\alpha_i . v$ belongs to $D_{\tilde{T}} \cap D_S$ when v belongs to $D_{\tilde{T}^\alpha_N} \cap D_{S^\alpha_N}$.

Then (3) gives the following inequality :

(4) $\|(\tilde{J}_N^{\alpha})(v)\|^2 \leq \|(\tilde{T}_N^{\alpha})(v)\|^2 + \|S_N^{\alpha}(v)\|$, all $v \in D_{\tilde{T}_N^{\alpha}} \cap D_{S_N^{\alpha}}$.

Now the announced lemma is provided by theorem 8.3.

When (4) is satisfied and $S_N^{\alpha} \circ T_N^{\alpha} = 0$, the equation $T_N^{\alpha}(u) = v$ has a unique solution in $D_{\tilde{T}_N^{\alpha}} \cap \overline{R_{T_N^{\alpha}}}$ for any $v \in R_{T_N^{\alpha}} \cap \operatorname{Ker} S_N^{\alpha}$ and u verifies :

$$\|u\|^2 \leq \|(\tilde{J}_N^{\alpha})(v)\|^2 .$$

§ 2.- A Runge theorem

Let E be a c. v. s. whose points are denoted by y . Let A be a complex manifold X be given, $\mathcal{P}(X,E)$ is the set of functions $f(x,y)$ defined on $X \times E$ such that $x \to f(x,y)$ belongs to $\mathcal{O}(X)$ and $y \to f(x,y)$ is a polynomial.

Theorem 8.5.- *Suppose X a finite dimensional Stein manifold, let K a holomorphically convex compact set in X and U a neighbourhood of K be given.*

Then there exists a relatively compact neighbourhood W of K such that

(i) $K \subset W \subset \bar{W} \subset U$

(ii) For any $f \in \mathcal{P}(U,E)$ and $\varepsilon > 0$, there exists $g \in \mathcal{P}(X,E)$ such that :

$$\|g(x,y) - f(x,y)\|_K \leq \varepsilon \, . \, \|f(x,y)\|_W \ , \text{ all } y \in E \ .$$

Furthermore, g can be chosen continuous in $X \times E$ whenever f is continuous in $X \times E$.

Proof.- Let ω be an open Stein sub-manifold of X , such that ω is relatively compact and contains K . To begin with, we construct g in $\mathcal{P}(\omega,E)$.

Since K is holomorphically convex, there exists $N \in \mathbb{N}$, $g \in \mathcal{O}(X)^N$ such that $W = g^{-1}(\Delta) \cap U$ is a relatively compact open set which verifies $K \subset W \subset \bar{W} \subset U$; here Δ is the unit polydisk of $\mathbb{C}^N$ whose points are denoted by z . For all ρ $]0,1[$ the set $M = g^{-1}(\overline{\rho.\Delta}) \cap W$ is compact and ρ can be chosen such that K is contained in M . Let $\chi \in C^{\infty}(\omega)$, with support in W and values 1 in a neighbourhood of M . Firstly, y is fixed in E . We apply the lemma 8.1 on the Stein manifold

$\omega \times \rho.\Delta$; since $\bar{d}\chi \cdot f \cdot \left[\Sigma\ |z_i - g_i(x)|^2\right]^{-1/2}$ belongs to $C^\infty(W \times \rho\Delta)$, the previous lemma gives a unique solution of the equation $-\bar{d}\chi \cdot f = \Sigma\left[z_i - g_i(x)\right]\ \bar{d}u_i$ in a convenient space $\left[L_o^2(\varphi)\right]^2$

$$(5) \qquad \|u\|^2 \leq \int |\bar{d}\chi \cdot f|^2\ \Sigma\ |z_i - g_i(x)|^2 \cdot e^{-\varphi}\ d\sigma \quad .$$

For this solution the function $G = \chi \cdot f + \Sigma\left[z_i - g_i(x)\right] u_i$ verifies $\bar{d}G = 0$ and so is analytic in $\omega \times \rho.\Delta$.

Now, the uniqueness of solutions implies that u is polynomial like f when y describes E and thus G belongs to $\mathcal{P}(\omega \times \rho\Delta, E)$.

Using mean integral values and (5), we get the following estimations for G :

$$(6) \qquad |G(x,z,y)|^2 \leq k_1(x,z) \int |G(x,z,y)|^2\ e^{-\varphi}\ d\sigma$$

$$(7) \qquad |G(x,z,y)| \leq k_2(x,z) \quad . \quad \|f(x,y)\|_W \quad .$$

Here, k_1 and k_2 are functions which are locally bounded in $\omega \times \rho.\Delta$.

The Taylor expension of G in $\rho.\Delta$ gives $G_n(x,y,z) = \frac{1}{2\pi i}\int G(x,ze^{i\theta},y)\ e^{-ni\theta}\ d\theta$ for all $z \in \rho.\Delta$ and G_n belongs to $\mathcal{P}(\omega, \mathbb{C}^N \times E)$. Since $g(K)$ is a compact set of $\rho.\Delta$, there exists $\lambda > 1$ such that $\lambda.g(K)$ is contained in $\rho.\Delta$ and we get by (7) :

$$(8) \quad \|G(x,g(x),y) - \sum_{n \leq q} G_n(x,g(x),y)\|_K \leq$$

$$(\lambda^{q+1} - \lambda^q)^{-1} \cdot \|k_2(x,g(x))\|_K \circ \|f(x,y)\|_W$$

Pointing out the equality of $G(x,g(x),y)$ and f in K and $G_n(x,g(x),y)$ belonging to (ω,E) , the inequality (1) is verified in ω , with q sufficiently large and $g(x,y) = \sum_{n \leq q} G_n(x,g(x),y)$.

Moreover the Cauchy integral for G_n and the estimation (7) give the following estimation for g

$$|g(x,y)| \leq \left(\sum_{n \leq q} \left(\frac{2}{\rho}\right)^n |g(x)|^n\right) \cdot \|k_2(x,z)\|_{1/2.\rho.\Delta} \cdot \|f(x,y)\|_W \ .$$

Then there exists a locally bounded function in ω such that :

(9) $|g(x,y)| \leqslant k_3(x) \, . \, \|f(x,y)\|_W$, in $\omega \times E$.

Second step.- We extend (1) in the whole X . There exists a sequence of open sets ω_n such that $\omega_n \subset \overline{\omega}_n \subset \hat{\omega}_n \subset \omega_{n+1}$, ω_n is an open Stein sub-manifold of X , $X = \bigcup \omega_n$, K is contained in ω_1 and $f \in \mathcal{P}(\omega_2,E)$. By induction, we are constructing a sequence f_n such that $f_1 = f$ and :

(i) $f_n \in \mathcal{P}(\omega_{n+1},E)$ and $\|f_{n+1}(x,y) - f_n(x,y)\|_{\hat{\omega}_n} \leqslant \varepsilon \, . \, 2^{-n} \, \|f(x,y)\|_W$

(ii) $|f_{n+1}(x,y)| \leqslant k_n(x) \, \|f(x,y)\|_W$, all $x \in \omega_{n+2}$

(iii) W is a suitable neighbourhood of $\hat{\omega}_1$, and each k_n is locally bounded in ω_{n+2}

Actually the first step gives f_2 and W . Now suppose $f_2,\ldots,f_{n+1}$ be constructed. Then, the first step gives $f_{n+2} \in \mathcal{P}(\omega_{n+3},E)$ such that $\|f_{n+2}(x,y) - f_{n+1}(x,y)\|_{\hat{\omega}_{n+1}} \leqslant \eta \, . \, \|f_{n+1}(x,y)\|_{W_n}$, here W_n is a set contained in ω_{n+2} . The estimation (ii) at the nth order gives (i) at the $(n+1)^{th}$ order for η small enough. By (9) we get the following :

$|f_{n+2}(x,y)| \leqslant k(x) \, \|f_{n+1}(x,y)\|_{W_n}$, with k locally bounded in ω_{n+3} ; then after using (ii) at the nth order we check (ii) at the $(n+1)^{th}$ order.

Lastly, it is clear by (i) that $f_n(x,y)$ is uniformly convergent on any compact set of X . Then its limit g belongs to $\mathcal{P}(X,E)$ and satisfies the inequality (1) .

Third step.- If f is locally bounded in $U \times E$, each f_n is also locally bounded in $\omega_{n+1} \times E$ by (ii). Furthermore (ii) provides $\|g(x,y) - f_n(x,y)\|_{\hat{\omega}_n} \leqslant 2 \, . \, \varepsilon \, . \, 2^{-n} \, \|f(x,y)\|_W$; thus g is locally bounded in $\omega_n \times E$ for each n and g is continuous with f together .

Corollary 8.6.- For any open set ω in X , we denote by $\mathcal{P}_c(\omega,E)$ the set of functions which belongs to $\mathcal{P}(\omega,E)$ and is continuous in $\omega \times E$.

If ω is an open Stein sub-manifold of X , $\mathcal{P}_c(X,E)$ is sequentially *dense in* $\mathcal{P}_c(\omega,E)_b$.

Proof.- Let K_n be an increasing sequence of compact holomorphically convex sets whose union is ω , and $f \in \mathcal{P}_c(\omega,E)$ be given. By theorem 8.2 there exists a relatively compact set W_n in ω such that for any $\varepsilon > 0$ there exists $g \in \mathcal{P}_c(X,E)$ which satisfies

$$\|f(x,y) - g(x,y)\|_{K_n} < \varepsilon \cdot \|f(x,y)\|_{W_n} .$$

Since f is continuous there exists a neighbourhood U_n of the origin of E such that f is bounded in $W_n \times U_n$. We choose $\varepsilon = 2^{-n} \cdot \|f\|^{-1}_{W_n \times U_n}$; that provides a sequence $f_n \in \mathcal{P}_c(X,E)$ such that $2^n(f - f_n)$ is bounded on any compact set of $\omega \times E$. Hence the sequence f_n is convergent to f in $\mathcal{P}_c(\omega,E)_b$.

Corollary 8.7.- Let K be a compact set in X , we introduce the space $\mathcal{P}_c(K,E)_b = \varinjlim \mathcal{P}_c(\omega,E)_b$ when ω is running through the neighbourhoods of K .

If K is holomorphically convex, $\mathcal{P}_c(X,E)$ is sequentially dense in $\mathcal{P}_c(K,E)_b$.

Proof.- It is an obvious consequence of the corollary 8.6 after pointing out a fundamental system of Stein neighbourhoods of K .

Definition 8.1.- We assume E is a complex c. v. s. An open set Ω in $X \times E$ is X-equilibrated iff $X \times \{0\}$ is contained in Ω and $(x,y) \in \Omega$ implies $(x,t.y) \in \Omega$, all $|t| \leqslant 1$.

Theorem 8.8.Let X be afinite dimensional Stein manifold, Ω a X-equilibrated open set in $X \times E$, ω an open Stein sub-manifold of X be given. Then $\mathcal{P}_c(X,E)$ is sequentially dense in $\mathcal{O}((\omega \times E) \cap \Omega)_b$.

Proof.- By corollary 8.6 , we have just to prove that $\mathcal{P}_c(\omega,E)$ is sequentially dense in $\mathcal{O}((\omega \times E) \cap \Omega)_b$. That is the next lemma.

Lemma 8.9.- Let X a complex manifold not necessarily finitely dimensional, Ω a X-equilibrated open set in $X \times E$ be given. Then $\mathcal{P}_c(X,E)$ is sequentially dense in $\mathcal{O}(\Omega)_b$.

Proof.- Let $f \in \mathcal{O}(\Omega)$ be given ; we use the Taylor expension $f = \sum_{n \geqslant 0} f_n$

with $f_n(x,y) = \frac{1}{2\pi i} \int f(x, e^{i\theta} . y)\, e^{-ni\theta}\, d\theta$ for all $(x,y) \in \Omega$. Clearly f_n belongs to $\mathcal{P}_c(X,E)$. Let ε_n be a sequence which converges to 1 and $(\varepsilon_n)^n$ converges to infinity. For any compact Q in Ω , $\tilde{Q} = \{(x,ty) \mid (x,y) \in Q , |t| < 1\}$ is another compact set in Ω and for a suitable $\rho > 1$ we get :

$$\varepsilon_N^N \| f - \sum_{n \leq N} f_n \|_Q \leq \left(\frac{\varepsilon_N}{\rho}\right)^N \frac{1}{\rho - 1} \| f \|_{\rho . \tilde{Q}} .$$

Then the sequence $\varepsilon_N^N (f - \sum_{n \leq N} f_n)$ is bounded on any compact set of Ω , therefore $\sum_{n \leq N} f_n$ is convergent to f in $\mathcal{O}(\Omega)_b$.

<u>Theorem</u> *8.9.- Let* X *be a manifold spread over a Frechet c. v. s.* E *and suppose* X *be the maximal extension of a s. i. d., natural Frechet space in* $\mathcal{O}(X)$*. Let* F *be another sequentially complete c. v. s. and* Ω *an* X*-equilibrated domain in* $X \times F$ *together be given.*

Then the $\mathcal{O}(\Omega)$*-maximal extension for* $\mathcal{O}_{E \times F}(\mathbb{C})$ *is an open set of* $X \times F$ *.*

<u>Proof</u>.- Let $\tilde{\Omega}$ be the domain of $X \times F$ which is maximal among the open sets of $X \times F$ which extend Ω for $\mathcal{O}_{E \times F}(\mathbb{C})$. Let $u : \tilde{\Omega} \to (Y,q)$ be the maximal extension of $\tilde{\Omega}$ for $\mathcal{O}_{E \times F}(\mathbb{C})$.

The analytic mappings $\varphi : (x,a) \to x$ from $\tilde{\Omega}$ to X and $\psi : (x,a) \to a$ from $\tilde{\Omega}$ to F can be extended according to $\tilde{\varphi}$ (resp. $\tilde{\psi}$) in $\mathcal{O}_Y(Y,X)$ (resp. $\mathcal{O}_Y(Y,F)$ by theorems 7.6 and 7.1. Let us consider the mapping $v : y \to (\tilde{\varphi}(y), \tilde{\psi}(y))$ from Y into $X \times F$, that verifies $v \circ u$ = identity and $p \circ v \circ u = q \circ u = p$, here p is the projection of $X \times F$ into $E \times F$. The the equality of $p \circ v$ and q on $u(\tilde{\Omega})$ entails the equality everywhere and v is a morphism from Y into $X \times F$.

Now, we prove that v is one-to-one, that will entail that Y is an open set of $X \times F$ (up to an isomorphism) which contains $\tilde{\Omega}$ and the equality $\tilde{\Omega} = Y$ since $\tilde{\Omega}$ is maximal.

Actually, let y' and y'' be given in Y such that $v(y') = v(y'')$. For any $f \in \mathcal{O}(\tilde{\Omega})$, the nth derivative $f^n(x,a)$ belongs to $\mathcal{P}_c(X,F)$ and has an extension g_n in $\mathcal{O}(Y)$ which can be written $g_n = f^n \circ v$; thus we get $g_n(y') = g_n(y'')$. The extension g of f in $\mathcal{O}(Y)$ verifies the following relation by theorem 6.12 and lemma 8.9

$$g(y') = \lim . g_n(y') = \lim g_n(y'') = g(y'') .$$

Since Y is separated by $\mathcal{O}(Y)$ we have $y' = y''$ and the proof is complete.

When E is finite dimensional and F is a Frechet space the previous result is improved in the following manner.

<u>*Corollary*</u> *8.9.- Let X a manifold spread over a finite dimensional E and Ω a X-equilibrated domain in $X \times F$ be given. Then the $\mathcal{O}(\Omega)$-maximal extension for $\mathcal{O}_{E\times F}(\mathbb{C})$ is an open set $\tilde{\Omega}$ in $\tilde{X} \times F$ with $\tilde{X}$ as the $\mathcal{O}(X)$-maximal extension of X.*

<u>Proof</u>.- We have only to verify that $\tilde{\Omega}$ being constructed in $\tilde{X} \times F$ as in theorem 8.9 contains $\tilde{X}$ and after that apply this theorem.

By contradiction, we take a compact set $\tilde{K}$ in $\tilde{X}/\tilde{\Omega}$. For any $f \in \mathcal{O}(\tilde{\Omega})$, we denote by $g_n(a)$ the extension of the mapping $x \to f^n(x,a)$ to $\mathcal{O}(\tilde{X})$. There exists a compact set K in X such that $\|g_n(a)\|_K \leqslant \|f^n(a)\|_{\tilde{K}}$, all $f \in \mathcal{O}(\tilde{\Omega})$, all $a \in F$. There exists a balanced neighbourhood ω of the origin in F such that $K \times \omega \subset \Omega$. For any balanced compact set Q in ω we have $\|g_n\|_{\tilde{K}\times Q} < \|f\|_{K\times Q}$.

The last inegality entails the summability of the series $\Sigma\, g_n(x,a)$ on $K \times \omega$, its sum belongs to $\mathcal{O}(K \times U)$ and is an extension of f. The maximality of $\tilde{\Omega}$ gives the announced result.

<u>Comment</u>.- By the same way it is possible to extend the corollary 8.9 to infinite dimensional E if the Nachbin's topology [62] is bornological. But very few results are known [18,29] and nothing concerning this problem is known on manifolds.

Index of Symbols

$\mathbb{N}$	set of integers
$\mathbb{R}$ (resp. $\mathbb{C}$)	real (resp. complex) field
(X,p)	spread manifold
$C_E(Z)$	presheaf of continuous and Z-valued germs on E
$F_E(U,Z)$	section over U of a sub-presheaf $F_E(Z)$ of $C_E(Z)$
$\mathcal{O}_E(Z)$	sub-presheaf of $C_E(Z)$ provided by complex analytic germs
$\mathcal{O}^b_E(Z)$	sub-presheaf of $\mathcal{O}_E(Z)$ provided by locally bounded analytic germs
$A_E(Z)$	sub-presheaf of $C_E(Z)$ provided by real analytic germs
$\mathcal{O}(\omega)$	$= \mathcal{O}_E(\omega, \mathbb{C})$
c. v. s.	Hausdorff locally convex vector space
s. i. d.	strongly invariant by derivation defined in section IV,1
sp.	topological spectrum
$sp.^{*}$, $sp.^{**}$	defined in section VI,2
$S(A)$, $\tilde{X}(A)$	defined in section VI,1
$N(E)$	set of continuous semi-norms on a c. v. s. E
E'	topological adjoint space of E
$L(E,Z)$	continuous endomorphisms from the c. v. s. E to the other Z
$L(E)$	$= L(E,E)$
E'_β (resp. E'_σ)	E' equipped with strong (resp. weak) topology
E'_Γ	E' equipped with the topology induced by a Frechet sub-space of $\mathcal{O}(\omega)$ which contains E'
$(.)_{c,resp.pc,resp.s}$	(.) equipped with compact (resp. precompact) (resp. pointwise) topology
$(.)_\varepsilon$, $(.)_b$	(.) equipped with ε-(resp. b) topology defined in section VI,2
$f^n(a)$	n^{th} derivation of f
$f^n(x,a)$	n^{th} derivation of f at x
$h(a)$	homomorphism h translated to a , defined in section VI,1
$\mathfrak{F}(I)$	set of finite subsets of I

$\ell_\infty(I)$	set of bounded mappings from I to $\mathbb{C}$ equipped with the uniform norm
$c_{o,o}$	subspace of $\ell_\infty(\mathbb{N})$ provided by sequences which vanish for n large enough
$K(X)$	set of convex, balanced, equicontinuous, precompact parts of $\mathcal{O}(X)_s$
$\hat{T}(A)$	A-convex hull of T
$\|f\|_T$	sup $\|f(x)\|$, when x describes T
d_X^v	defined in section IV,4
D_A (resp. R_A)	domain (resp. range) of the linear operator T transposed operator of A
$\tilde{A}$	transposed operator of A
$\mathcal{P}(X,E)$, $\mathcal{P}_c(X,E)$	mappings from $X \times E$ to , analytic on X and polynomial on E , (resp. continuous mappings...)
$\otimes_\pi$	tensor product equipped with projective topology
card.	cardinal of ...
$\aleph_0$	card. $\mathbb{N}$

BIBLIOGRAPHY

[1] H. ALEXANDER Analytic functions on Banach spaces
Univ. Berkley (1968)

[2] R.M. ARON The bornological topology on the space of holomorphic mappings on a Banach space
C.R. Ac. Sc. Paris (t. 272) (1971)

[3] V. AURICH Spectrum as envelope of holomorphy
Publ. Univ. Münschen (1973)

[4] J.A. BARROSO and L. NACHBIN Sur certaines propriétés bornologiques des espaces d'applications holomorphes
Colloque de Liège (1970)

[5] J.A. BARROSO Topologias nos espaços de applicaços holomorphas entre espaços localmente convexos
Publ. Univ. Rio de Janeiro (1970)

[6] J. BOCHNAK Analytic functions in Banach spaces
Studia Math. t. 35 (1970)

[7] J. BOCHNAK and J. SICIAK
a) Polynomial and multilinear mappings in topological vector spaces
b) Analytic functions in topological vector spaces
Studia Math. t. 39 (1971)

[8] S. BOCHNER and W.T. MARTIN Several complex variables
Princeton Univ. Press. (1948)

[9] P.J. BOLAND Some spaces of entire and nuclearly entire functions on Banach space
C.R. Ac. Sc.

[10] H.J. BREMERMANN Complex convexity
Trans. Ams. t. 82 (1956)

[11] H.J. BREMERMANN Holomorphic functionnals and complex convexity in Banach spaces
Pac. J. Math. t. 7 (1957)

[12] " The envelope of holomorphy of tube domains in Banach spaces
Pac. J. Math. t. 10 (1960)

[13] " Pseudo-convex domains in t. v. s.
Proc. Univ. Mineapolis Springer (1965)

[14] BUCHWALTER Topologies et compactologies
Pub. Univ. Lyon (1969)

[15] H. CARTAN Sem. E.N.S. 51/52
New-York Benjamin (reprint 1967)

[16] S.B. CHAE Holomorphic germs on Banach spaces
Ann. Inst. Fourier t. 21 (1971)

[17] G. COEURÉ Fonctions plurisousharmoniques sur les espaces vectoriels topologiques et applications à l'étude des fonctions analytiques
Ann. Inst. Fourier t. 20 (1970)

[18] " Fonctionnelles analytiques sur certains espaces de Banach
Ann. Inst. Fourier t. 21 (1971)

[19] " Fonctions plurisousharmoniques et analytiques à une infinité de variables
C.R. Ac. Sc. Paris t. 267 (1968)

[20] " Le théorème de convergence sur les espaces localement convexes complexes
C.R. Ac. Sc. Paris t. 264 (1967)

[21] J.F. COLOMBEAU Exemples d'applications G-analytique, analytique, différentiable en dimension infinie
C.R. Ac. Sc. Paris t. 273 (1971)

[22] J.F. COLOMBEAU and D. LAZET — Sur les théorèmes de Vitali et Montel en dimension infinie
C.R. Ac. Sc. Paris t. 274 (1972)

[23] S. DINEEN — Holomorphy types on a Banach space
Studia Math. t. 39 (1971)

[24] " — Bounding subsets of a Banach space
Math. Ann. t. 92 (1971)

[25] " — The Cartan-Thullen theorem for Banach spaces
Ann. Sc. Norm. (Pisa) t. 24 (1970)

[26] " — Unbounded holomorphic functions on a Banach space
J. London Math. Soc. t. 4 (1971)

[27] " — Holomorphic functions on locally convex topological vector spaces
C.R. Ac. Sc. Paris t. 274 (1972)

[28] " — Holomorphically complete locally convex topological vector spaces
Sem. P. Lelong, LectureNotes : 332 (1972)

[29] " — Holomorphic functions on (C_o,X_o)-modules
Math. Ann. t. 196 (1972)

[30] " — Runge domains in Banach spaces
Proc. Roy. Irish Acad. t. 71 (1971)

[31] S. DINEEN and A. HIRSCHOWITZ — Sur le théorème de Levi Banachique
C.R. Ac. Sc. Paris t. 272 (1971)

[32] A. DOUADY — Le problème des modules pour les sous-espaces analytiques compacts d'un espace analytique donné
Ann. Inst. Fourier t. 16 (1966)

[33] T.A.W. DWYER — Partial differential equations in Fischer Fock spaces
Bull. Ams. t. 77 (1971)

[34] S.J. GREENFIELD The Hilbert Ball and bi-ball are holomorphically inequivalent
Bull. Ams. t. 77 (1971)

[35] " Automorphism groups of bounded domains in Banach spaces
Trans. of the Ams. t. 166 (1972)

[36] L. GRUMAN and C.O. KISELMAN Le problème de Levi dans les espaces de Banach à base
C.R. Ac. Sc. Paris t. 274 (1972)

[37] L. GRUMAN The Levi problem in certain infinite dimensional vector spaces (to publish)

[38] C. GUNNING and H. ROSSI Analytic functions of several complex variables
Prentice - Hall Inc. (1965)

[39] D.P. GUPTA Malgrange theorem for nuclearly entire functions of bounded type on Banach space
Notas de Mat. n° 37 (1968)

[40] M. HERVÉ Analytic and plurisubharmonic functions
Springer lecture notes : 198 (1970)

[41] Y. HERVIER Sur le problème de Levi pour les espaces étalés banachiques
C.R. Ac. Sc. Paris t. 275 (1972)

[42] " On the Weierstrass Problem in Banach spaces
Coll. Lexington (1973)

[43] E. HILLE and E.G. PHILLIPS Functionnal analysis semi-groups
Coll. Ams. (1957)

[44] A. HIRSCHOWITZ Sur les suites de fonctions analytiques
Ann. Inst. Fourier t. 20 (1970)

[45] " Remarques sur les ouverts d'holomorphie d'un produit dénombrable de droites
Ann. Inst. Fourier t. 19 (1969)

[46] A. HIRSCHOWITZ Sur un théorème de M.A. Zorn
Publ. Univ. Nice (1970)

[47] " Sur le non-prolongement des variétés analytiques banachiques réelles
C.R. Ac. Sc. Paris t. 269 (1969)

[48] " Prolongement analytique en dimension infinie
Ann. Inst. Fourier t. 22 (1972)

[49] H. HOGHE-NLEND Deux remarques sur les applications analytiques en dimension infinie
Ann. Ac. Bras. Ciencias

[50] L. HÖRMANDER An introduction to complex analysis in several variables
Van Nostrand (1966), North-Holland (1973)

[51] J. HORVATH Topological vector spaces and distributions
Addison - Wesley (1966)

[52] JOSEPHSON Pseudo-convex and holomorphically convex domains
Upsala Univ. (not published)

[53] C.O. KIESELMAN On entire functions of exponential type and indicators of analytic functionals
Acta Math. t. 117 (1967)

[54] " Plurisubharmonic functions in vector spaces
Upsala Univ. (not published)

[55] P. LELONG Les fonctions plurisousharmoniques
Ann. E.N.S. t. 62 (1945)

[56] " Fonctions plurisousharmoniques dans les espaces vectoriels topologiques
Sem. Lelong Springer lectures Notes n° 71, 116, 205

[57] " Applications analytiques et théorème de Banach Steinhaus polynomial

[58] P. LELONG Fonctions entières et fonctionnelles analytiques
Sem. Montréal (1967)

[59] " Fonctions plurisousharmoniques et ensembles polaires dans les espaces vectoriels topologiques
C.R. Ac. Sc. Paris t. 267 (1968)

[60] M.C. MATOS Domains of τ-holomorphy in separables Banach spaces
Math. Ann.

[61] " Sur l'enveloppe d'holomorphie des domaines de Riemann sur un produit dénombrable de droites
C.R. Ac. Sc. Paris t. 271 (1970)

[62] L. NACHBIN Topology on spaces of holomorphic mappings
Springer verlag (1969)

[63] " Holomorphic functions, domains of holomorphy, local properties
North-Holland (1970)

[64] " Sur les espaces vectoriels topologiques d'applications continues
C.R. Ac. Sc. Paris t. 271 (1970)

[65] " Concerning holomorphy type for Banach spaces
Studia Mathematica t. 38 (1970)

[66] " Convolution operators in spaces of nuclearly entire functions on a Banach space
Functionnal Analysis and related fields
Ed. Browder (1970)

[67] PH. NOVERRAZ Sur la pseudo-convexité et la convexité polynomiale en dimension infinie
C.R. Ac. Sc. t. 274 (1972)

[68] " Fonctions plurisousharmoniques et analytiques dans les espaces vectoriels topologiques
Ann. Inst. Fourier t. 19 (1969)

[69] PH. NOVERRAZ Prolongement, completion pseudo-convexe et approximation en dimension infinie
C.R. Ac. Sc. Paris t. 276 (1973)

[70] " Pseudo-convexité, convexité polynomiale et domaine d'holomorphie en dimension infinie
North-Holland (1973)

[71] D. PISANELLI Applications analytiques en dimension infinie
C.R. Ac. Sc. Paris t. 274 (1972)

[72] J.P. RAMIS Sous-ensembles analytiques d'une variété banachique complexe
Springer (1970)

[73] C.E. RICKART Holomorphic convexity for general function algebra
Can. J. of math. t. 20 (1968)

[74] " Analytic phenomena in general function algebra
Pacific J. of math. t. 18 (1966)

[75] " Analytic functions of an infinite number of complex variables
Duke math. Jour. t. 36 (1969)

[76] " Plurisubharmonic functions and convexity properties for general functions algebra
Trans. of the Ams. t. 169 (1972)

[77] H.H. SCHAEFFER Topological vector spaces
Springer (1970)

[78] M. SCHOTTENLOHER Holomorphe vervollständigun metrizierbar lokal konvexer Räume
Sitz Bayer Akad Wiss (1973)

[79] " Uber analytischer fortsetzung in Banachraumen
Math. Ann. t. 199 (1972)

[80] " Analytic continuation and domains spread
Proc. conf. Dublin (1973)

[81] F. SERGERARD — Un théorème des fonctions implicites sur certains espaces de Fréchet
Ann. Sc. Ec. N. Sup. Paris t. 5 (1972)

[82] K. STEIN — Einführung in die functionentheorie meherer veränderlichen Vorlesungsausarbeitung
München (1962)

[83] L. WAELBROECK — Topological vector spaces and algebras
Springer lectures notes n° 230

[84] M.A. ZORN — Characterisation of analytic functions in Banach spaces
Ann. of Math. t. 46 (1945)

[85] " — Gateaux differentiability and essential boundness
Duke math. Jour. t. 12 (1945)

8480-21
5-01